EMOTIONS, TECHNOLOGY, AND DESIGN

Emotions and Technology
Communication of Feelings for, with, and through Digital Media

Series Editor
Sharon Y. Tettegah

Emotions, Technology, and Design

Volume Editors
Sharon Y. Tettegah and Safiya Umoja Noble

EMOTIONS, TECHNOLOGY, AND DESIGN

Edited by

SHARON Y. TETTEGAH
Professor, University of Nevada, Las Vegas
College of Education, Las Vegas, NV, USA

Beckman Institute for Advanced Science and Technology,
National Center for Supercomputing Applications, affiliate,
University of Illinois, Urbana, IL, USA

SAFIYA UMOJA NOBLE
Department of Information Studies, University of California,
Los Angeles (UCLA), Los Angeles, CA, USA

AMSTERDAM • BOSTON • HEIDELBERG • LONDON
NEW YORK • OXFORD • PARIS • SAN DIEGO
SAN FRANCISCO • SINGAPORE • SYDNEY • TOKYO
Academic Press is an imprint of Elsevier

Academic Press is an imprint of Elsevier
125 London Wall, London, EC2Y 5AS, UK
525 B Street, Suite 1800, San Diego, CA 92101-4495, USA
50 Hampshire Street, 5th Floor, Cambridge, MA 02139, USA
The Boulevard, Langford Lane, Kidlington, Oxford OX5 1GB, UK

© 2016 Elsevier Inc. All rights reserved.

No part of this publication may be reproduced or transmitted in any form or by any means, electronic or mechanical, including photocopying, recording, or any information storage and retrieval system, without permission in writing from the publisher. Details on how to seek permission, further information about the Publisher's permissions policies and our arrangements with organizations such as the Copyright Clearance Center and the Copyright Licensing Agency, can be found at our website: www.elsevier.com/permissions.

This book and the individual contributions contained in it are protected under copyright by the Publisher (other than as may be noted herein).

Notices

Knowledge and best practice in this field are constantly changing. As new research and experience broaden our understanding, changes in research methods, professional practices, or medical treatment may become necessary.

Practitioners and researchers must always rely on their own experience and knowledge in evaluating and using any information, methods, compounds, or experiments described herein. In using such information or methods they should be mindful of their own safety and the safety of others, including parties for whom they have a professional responsibility.

To the fullest extent of the law, neither the Publisher nor the authors, contributors, or editors, assume any liability for any injury and/or damage to persons or property as a matter of products liability, negligence or otherwise, or from any use or operation of any methods, products, instructions, or ideas contained in the material herein.

Library of Congress Cataloging-in-Publication Data
A catalog record for this book is available from the Library of Congress

British Library Cataloguing in Publication Data
A catalogue record for this book is available from the British Library

ISBN: 978-0-12-801872-9

For information on all Academic Press publications
visit our website at http://store.elsevier.com/

Publisher: Nikki Levy
Acquisition Editor: Emily Ekle
Editorial Project Manager: Timothy Bennett
Production Project Manager: Caroline Johnson
Designer: Matthew Limbert

Typeset by SPi Global, India

Printed and bound in the United States of America

CONTENTS

Contributors — ix
Foreword — xi
Preface — xv

Section I Experiments and Theories in Emotions, Technology, and Design — 1

1. Emotional Screen: Color and Moving Images in Digital Media — 3
Federico Pierotti

Color as Cinematic Emotion: A Historical Summary — 4
From Brain to Marketing: The Attraction Effect — 6
Patterns of Emotion: Three Examples from Contemporary Digital Media — 9
Concluding Remarks — 15
References — 16

2. Safe and Sound: Using Audio to Communicate Comfort, Safety, and Familiarity in Digital Media — 19
Michael L. Austin

Hearing, Listening, Feeling — 21
(Auditory) Displays of Emotion — 28
Conclusion: Sounding Safety, Security, Stasis, and Status — 33
References — 34

3. Emoticons in Business Communication: Is the :) Worth it? — 37
Jennifer M. Loglia, Clint A. Bowers

Introduction — 37
Nonverbal Communication — 39
Emoticons — 42
Leader-Member Exchange — 46
References — 50

4. Empathetic Technology — 55
Néna Roa Seïler, Paul Craig

Introduction — 55
Empathy and Technology — 57

Modeling Empathy	59
Empathetic Virtual Companions	63
Perceptions of Empathetic Virtual Agents	68
Developing an Interaction Strategy for a Virtual Companion	76
References	77

5. Spoken Dialog Agent Applications using Emotional Expressions   83
Kaoru Sumi

Introduction	83
Experiment Investigating Impressions and Behavior Change Caused by Replies from the Agent	85
Comparative Experiment on Effects of a Virtual Agent and a Robot	88
A Spoken Agent System for Learning Customer Services	93
A Spoken Agent System for Mental Care using Expressive Facial Expressions and Positive Psychology	96
Discussion	100
Conclusion	101
Acknowledgments	101
References	102

6. Engaging Learners Through Rational Design of Multisensory Effects   103
Debbie Denise Reese, Dianne T.V. Pawluk, Curtis R. Taylor

Drawing on Two Process Models for Representation	105
Toward Rational Design Guidelines	116
Increased Interest and Engagement	121
The Research Agenda	122
References	123

7. Designing Interaction Strategies for Companions Interacting with Children   129
Néna Roa Seïler

Introduction	129
Agents as Companions	129
Companions, Children and Companionship	130
Current Companion Projects	131
The Affective Channel: A Framework for Emotionally Intelligent Companion Interaction	132
Designing Interaction Strategies for Companions	132
Emotional Interaction Strategies	137

Conversational Interaction Strategies (CIS)	137
Domain Specific Strategy	138
Wizard of Oz Experiment	140
The Experiment	142
The Subjects of the Experiment	142
The Game	143
The Wizard of Oz (WoZ) of Samuela, Nao and Ari	145
The Pilot Session	148
The Focus Groups and Interviews	149
The Implementation of the Experiment	149
Report of Findings	155
Personality Descriptors	156
Physical Descriptors	157
Utility or Functionality Descriptors	158
Interaction with a Companion	159
Preferred Activities with Companions	160
Preferred Companion for Supporting Different Subjects	160
Significance of Companions	161
Trust in Companions	161
Emotions toward Companions	162
Discussion of Results	163
Summary and Future Work	165
References	165

Section II Critical Theoretical Engagements with Emotions, Technology, and Design — 169

8. The Emulation of Emotions in Artificial Intelligence: Another Step into Anthropomorphism — 171
Mariana Goya-Martinez

Introduction: Redefining the Human	171
The Definition of Emotions in Artificial Intelligence	173
The Role of Emotions in Intelligent Agents' Design	177
Anthropomorphism in Artificial Intelligence	183
References	185

9. Through Google-Colored Glass(es): Design, Emotion, Class, and Wearables as Commodity and Control — 187
Safiya Umoja Noble, Sarah T. Roberts

Introduction: In the Google Gaze	187
Unexamined Occupation: San Francisco and the Crystallization of Glass and Class Rage	190

Google Glass's Panoptic Gaze: Surveillance and Emotion ... 193
Emotion and Resistance: The Emergence of the "Glasshole" and Public Pushback to Google Glass ... 197
Wearable Control ... 202
Analyzing Class through Glass ... 203
Surveying Emotions and Space through Design ... 205
Conclusion: The Future of Glass ... 208
References ... 210

10. Designing Emotions: *Deliver the Nets*, Eradicate Malaria in Africa, and Feel Good?   213
Ergin Bulut, Robert Mejia

Designing the Emotions of Communicative Capitalism ... 213
Serious Games Is Serious Work ... 214
Deliver the Nets: Eradicate Malaria, Reify Africa ... 217
Conclusion ... 223
References ... 224

11. Police Body Cameras: Emotional Mediation and the Economies of Visuality   227
Stacy Wood

Introduction ... 227
POV: Body-Worn Cameras in the Discourse(s) of Surveillance ... 228
Design Elements of an AntiEmotion ... 231
Paradoxical Space and the Politics of Visuality ... 236
Conclusion ... 237
References ... 238

Index ... 241

CONTRIBUTORS

Michael L. Austin
Howard University, Washington, DC, USA

Clint A. Bowers
University of Central Florida, Orlando, FL, USA

Ergin Bulut
Koc University, Istanbul, Turkey

Paul Craig
Xi'an Jiaotong-Liverpool University, Suzhou, China

Mariana Goya-Martinez
University of Illinois at Urbana-Champaign, Champaign, IL, USA

Jennifer M. Loglia
University of Central Florida, Orlando, FL, USA

Robert Mejia
SUNY Brockport, Brockport, NY, USA

Safiya Umoja Noble
Department of Information Studies, Graduate School of Education & Information Studies, University of California, Los Angeles, CA, USA

Dianne T.V. Pawluk
Virginia Commonwealth University, Richmond, VA, USA

Federico Pierotti
University of Florence, Firenze, Italy

Debbie Denise Reese
Zone Proxima, LLC, Wheeling, WV, USA

Sarah T. Roberts
Faculty of Information and Media Studies, The University of Western Ontario, London, ON, Canada

Néna Roa Seïler
School of Computing, Center for Interaction Design, Edinburgh Napier University, Edinburgh, UK

Kaoru Sumi
Future University Hakodate, Hakodate, Japan

Curtis R. Taylor
University of Florida, Gainesville, FL, USA

Stacy Wood
University of California, Los Angeles, CA, USA

FOREWORD

With respect to technology, it is important to place terms and tools within a historical context, given that in today's society when speaking to a person who is a Millennial (individuals who are born in the early 1980s to 2000), s(he) may tell you that technology is the Internet and Smart Phones. For the Millennial, then, technology may only mean digital or biotechnologies. If we were to speak broadly to some individuals from The Silent Generation, Boomers, Millennials, and Generation Y, technology may also mean automobiles, airlines, overhead projectors, flashlights, microwaves, ATMs, etc. Hence, technology in the twenty-first century can mean many things. For example, technology could mean software applications, hardware, social media platforms, functional magnetic resonance imaging, mobile technology, learning and content management systems, just to name a few.

Humans and other animals have used tools for centuries; however, the most important aspect of any tool is how we use and interact with it and the emotional responses we experience, while we interact with it either physically or psychologically. The focus of this book series is to provide a variety of conceptual, theoretical, and practical perspectives on the role of emotions and technology. Various psychological and social-emotional aspects of communicating through and with many types of technology are engaged in ways that extend our understanding of technology and its consequences on our lives.

A specific goal and purpose of this book series focuses on emotions and affective interactions with and through technology. In some cases, these interactions are user-to-user, supported by the technology. In other instances, these interactions are between the user and the technology itself. Let us take, for example, researchers who have used animated social simulation technology to measure emotions of educators (Tettegah, 2007) and others who use biotechnology to measure decision-making and emotional responses of users of technology (Baron-Cohen, 2011; Decety & Ickes, 2009). In a recent article, Solomon (2008) points out, "One of the most critical questions about human nature is the extent to which we can transcend our own biology" (p. 13). I would argue that through our use of technology we, in fact, are attempting to extend and transcend our emotions by way of robots and other intelligent technological agents. As such, we should then ask ourselves: why are discussions of emotions and technology so important?

Inquiry regarding the nature of emotions is not new. In fact, examples of such forms of inquiry have been documented since the dialogs of Socrates and Plato. Researchers and practitioners in psychology, sociology, education, and philosophy understand the complicated nature of emotions, as well as [the importance of] defining emotions and social interactions. The study of emotions is so complicated that we still continue to debate within the fields of philosophy, education, and the psychology, the nature of emotions and the roles of affective and cognitive processes involving human learning and behavior. The volumes in this series, therefore, seek to present important discussions, debates, and perspectives involving the interactions of emotions and various technologies. Specifically, through this book series on Emotions and Technology, we present chapters on emotional interactions with, from, and through technology.

The diversity of emotions played out by humans with and through technology run the gamut of emotions, including joy, anger, love, lust, empathy, compassion, jealousy, motivation, frustration, and hatred. These emotional interactions can occur through interactions with very human looking technologies (e.g., avatars, robots) or through everyday commonplace technologies (e.g., getting angry at an ATM machine when the user fails to follow directions). Hence, understanding the ways in which technology affords the mediation of emotions is extremely important toward enhancing our critical understanding of the ways in which student minds, through technology, are profoundly involved in learning, teaching, communicating, and developing social relationships in the twenty-first century.

The majority of the chapters presented in books included in the series will no doubt draw on some of the recent, pervasive, and ubiquitous technologies. Readers can expect to encounter chapters that present discussions involving emotions and mobile phones, iPads, digital games, simulations, MOOCs, social media, virtual reality therapies, and Web 2.0/3.0 technologies. However, the primary focus of this book series engages the readers in psychological, information communication, human computer interaction, and educational theories and concepts. In other words, technologies will showcase the interactions; however, the concepts discussed promise to be relevant and consistent constructs, whether engaging current technologies or contemplating future tools.

The book series began with a call for a single volume. However, there was such a huge response, that what was to be one volume turned into eight volumes. It was very exciting to see such an interest in the literature that lies at the intersection of emotions and technology. What is very clear here is that human beings are becoming more and more attached to digital technologies, in one form or another. In many ways, we could possibly posit the statement that many individuals in the world are inching their way toward becoming cyborgs. It is apparent that digital technologies are in fact more and more second nature to our everyday life. In fact, digital technologies are changing faster than we are aging.

The life of a new technology can be 6 months to 1 year, while human lifespan ranges from 0 to 80 years. With the aforementioned in mind, humans have to consider how their emotions will interact and interface with the many different technologies they will encounter over the course of a lifetime. It seems as if it were only yesterday when the personal computer was invented and now we have supercomputing on a desktop, billions of data at our fingertips on our smartphone computers, and nanotechnology assisting us with physiological functions of living human animals. Regardless of the technology we use and encounter, emotions will play a major role in personal and social activities.

The major role that technology plays can be observed through the many observations of how humans become excited, frustrated, or relieved when interacting with new technologies that assist us within our daily activities.

Our hope is that scholars and practitioners from diverse disciplines, such as Informatics, Psychology, Education, Computer Science, Sociology, Engineering and other Social Sciences and Science, Technology, Media Studies, and Humanities fields of study will find this series significant and informative to their conceptual, research, and educational practices. Each volume provides unique contributions about how we interact emotionally with, through, and from various digital technologies. Chapters in this series range from how intelligent agents evoke emotions, how humans interact emotionally with virtual weapons, how we learn or do not learn with technology, how organizations are using technology to understand health-related events, to how social media helps to display or shape our emotions and desires.

This series on Emotions and Technology includes the following volumes: (1) Emotions, Technology, and Games, (2) Emotions, Technology, Design, and Learning, (3) Emotions, Technology, and Behaviors, (4) Emotions, Technology, and Learning, (5) Emotions, Technology, and Health, (6) Emotions, Technology, and Design, (7) Emotions, Technology, and Social Media, and (8) Emotions and Mobile Technology.

Sharon Tettegah
University of Nevada, Las Vegas, USA

ACKNOWLEDGMENTS

I would like to give a special thank you to Martin Gartmeier, Dorothy Espelage, Richard Ferdig, WenHao David Huang, Grant Kien, Angela Benson, Michael McCreery, Safiya Umoja Noble, Y. Evie Garcia, and Antonia Darder and all of the authors for their reviews and contributions to this work.

REFERENCES

Baron-Cohen, S. (2011). *The science of evil*. New York: Basic Books.
Decety, J., & Ickes, W. (2009). In *The social neuroscience of empathy*. Cambridge: The MIT Press.
Solomon, R. C. (2008). The philosophy of emotions. In M. Lewis, J. M. Haviland-Jones, & L. F. Barrett (Eds.), *The handbook of emotions* (3rd ed., pp. 3–15). London: Guildford Press.
Tettegah, S. (2007). Pre-service teachers, victim empathy, and problem solving using animated narrative vignettes. *Technology, Instruction, Cognition and Learning, 5*, 41–68.

PREFACE

INTRODUCTION

Much has been written at the intersections of technology and design within human-computer interaction. Until recently, these works had little regard for affect or emotion as a primary driver in how people impact, and are impacted by, various digital technologies (Calvo, D'Mello, Gratch, & Kappas, 2015; Ivonin et al., 2015; see Tettegah & Huang, 2015; Tettegah & Gartmeier, 2015; Tettegah & Espelage, 2015 in this series). This volume highlights the importance of thinking about emotion as a key driver in the design of, and response to, various technological projects from a psychological, psycho-social, and social psychological perspective, including empathy and emotional attachment for artifacts (Kim & Ryu, 2014). Because design is both socio-political and instrumental, although not mutually exclusive, this book is broken into two sections that foreground these two features. Norman (1990) argued that design must account for multiple dimensions of emotion that span a continuum of processing at various stages of engagement in the design of technology. In many ways, the contributors to this volume are looking at multiple ways in which emotion is implicit in various forms of design associated with all types of technologies that go beyond human-computer interaction.

In the first section of the book, a number of concepts and experiments that demonstrate how emotion, technology, and design are constructed in service of a variety of human experiences are presented. Authors discuss psychological implications of how consumers interact with, acquire knowledge through, and consume using various technologies. The backgrounds of these contributions range from experimental psychology to the study of music and sound in the design of technologies, and the role emotion plays in human sense making. Taken together, these two sections of the book cover a broad range of concepts, theories, and empirical studies that deepen our understanding of the intricate interplay between human emotion and our technological engagements.

VOLUME OVERVIEW

Experiments and Theories in Emotions, Technology, and Design

In part one of this volume, experimental and theoretical studies at the intersection of emotions, technology, and design lead the readers to consider what aspects of design is important in consideration of emotions. The first chapter in this section, "Emotional Screen, Color, and Moving Images in Digital Media," by Federico Pierotti looks at the emotional impact of color in contemporary media display. Pierotti studies how cinema and neuroscience converge to evoke sensory impact through visuality and cognitive activity. By using three cases from media and cinema, he demonstrates the possibility for "building artificial and self-referential color patterns," that extend our thinking about the possibilities for an "emotional screen."

Michael L. Austin's research on digital media examines sound and its emotional evocation of comfort, safety, and familiarity. This chapter considers how sound summons emotion and communication in our engagements with digital media. By studying the ways that companies use sound to create strong, positive emotional attachments to products, Austin details the sound processes used in digital media, and the signifiers that "help to formulate, to various degrees, an emotional sense of self-worth" in purchasing experiences for consumers. He argues that sound has emotional impacts that must be understood as computing moves toward the "Internet of Things."

Following, Jennifer Loglia and Clint Bowers look closely at emoticons as a tool in the workplace as a means of clarifying the communication and reception of messages. Because computer mediated communications, specifically emails, are difficult to emotionally interpret, Loglia and Bowers argue that emoticons can play a valuable role in the workplace to convey nonverbal feelings and emotions, which can improve employee relationships.

Néna Roa Seïler and Paul Craig explore the intersections of empathy and technology, and how we can feel one another's emotional and cognitive experience. Seïler and Craig argue that empathy is essential to human communication, and that technologies must incorporate empathic features, particularly in technologies like virtual agents, in service of human relationships.

Kaoru Sumi's, "Spoken Dialog Agent Applications Using Emotional Expressions," is a multifaceted experimental study in emotions and virtual agents. Sumi's work is an experiment in how virtual agents can be of service to those who need mental health care, or hikikomori, which is the "the phenomenon of reclusive adolescents or young adults who withdraw from social life." Breakthroughs in the emotional efficacy of virtual agents, she argues, can support people who might suffer from bullying, post-traumatic stress disorder, suicidal tendencies, and other forms of social alienation and isolation. Sumi operationalizes a study that explores emotional interfaces in therapeutic technological agents.

Denise Debbie Reese, Diane T.V. Pawluk, and Curtis R. Taylor present research on multisensory representations in digital games, and how they can reinforce learning. In their work, they test game goals, analogous to learning goals, and how these assist learners to negotiate with "multisensory metaphor representations to discover and apply targeted relational structures." The goal of this and a future research agenda based on this study is to better understand how modalities affect learning, emotion, and interactions, specifically in the context of targeted learning.

To powerfully close this section, Néna Roa Seïler investigates the design of dynamic emotional interactions of companions with children. Companions are designed to go beyond embodied conversational agents that are endowed with various affect abilities. Ideally, companions are designed to provide total social emotional support for their caregivers. Roa Seïler stresses that a challenge for all application designers is to design and implement a set of Interaction Strategies for Companions that would afford a more organic process for the user.

Critical Theoretical Engagements with Emotions, Technology, and Design

In the second section of this book, we have grouped together contributors who are concerned with how technologies are designed, with an attendant quality of either producing or reproducing emotion. This section explores the politics of technological design (Pacey, 1983; Winner, 1986) and how design choices are loaded with a variety of social, political, and economic values that are never neutral or without consequence. These authors are largely concerned with the social, economic, ethical, and political factors at the intersections of emotion, technology, and design.

The section opens with Mariana Goya-Martinez, who explores the research trajectories of artificial intelligence and the humanization of digital technologies through anthropomorphization. This chapter investigates how emotions are defined and the role of emotion in a virtual agents' performance of humanness. Goya-Martinez has researched a range of virtual agent designs and their quests in improving empathic human-machine interaction, as well as the ways in which machines are increasingly being designed with cognitive processes that attempt to replicate rational thinking. Her work is primarily concerned with how emotions can be essential for creating systems of thought that can organize knowledge. The chapter also discusses the ethical dimensions of the technological body, including the "total anthropomorphism of the machine and a total machine-morphism of the human."

Safiya Umoja Noble and Sarah T. Roberts engage critical information theory to examine multiple dimensions of Google Glass, a wearable headpiece with an outward-facing camera that has induced severe emotional public backlash. In this chapter, they problematize wearable technologies like Google Glass as digital tools that are designed, ultimately, to profit from emotional data, including psychological and biological information. They posit Google Glass as a project loaded with design flaws that intrude upon both the physical and emotional domains of non-wearers. Such an exploration and occupation of public space, they argue, reflects disruptive and uncritically examined power relations by Google. These projects, they argue, are a manifestation of a class elitism that is part of the "design imaginary" of Google.

Following this, Ergin Bulut and Robert Mejia explore the "serious games" movement as an affective expression of "communicative capitalism." In this chapter, they suggest that serious games "transform complex social problems into enjoyable experiences that often work to naturalize the primacy of capitalism over alternative political economic systems." Through an analysis of the discourses of the game "Deliver the Nets," they demonstrate how playing games of this nature make the political conditions of poverty, and the economic and social policies of neoliberalism, virtually invisible. By engaging positive emotions and good feelings through serious games, multiple dimensions of global crises are rendered more difficult to intervene upon.

To close out the second section of this volume, authors engage critical theories of technology, emotion, and design. Stacy Wood presents a critical discussion about police body cameras and emotional mediation. This chapter explores the ways in which police body cameras are positioned as neutral agents that she says, "exist somehow outside of the exchange or incident itself as a tool for capture, as a neutral, anti-emotional, anti-biased technological object." Wood argues how the design of body cameras function differentially depending upon spatial and power relations that underlie surveillance technologies. Through an examination of the discourses of criminality, race, and victimization, Wood foregrounds the ways that body-worn surveillance cameras are designed; she discusses a myriad of attendant consequences that include volatility between police and victims of police brutality. Wood theorizes how "emotions become a part of an economy of visuality," through the apparatus of police body cameras.

Emotion, technology, and design are intertwined in multiple dimensions of sociality. The goal of this book is to explore a range of ideas, from critical perspectives to experimental design, in the convergence of and emergence of digital technologies that function and incorporate a variety of emotional conceptions. Technologies are never neutral, and, as narratives and engagements with them are increasingly studied, the human experiences of emotion are in constant interplay. The breadth and variety of chapters in this collection represent a wide range of research, which points to the importance of focusing on these three important fields of study. As editors of this volume, our goal is that this collection, in a series of volumes on emotions and technology, will further dialog and deepen engagement with a range of concerns presented herein.

Safiya Umoja Noble
University of California, Los Angeles, CA, USA

Sharon Tettegah
University of Nevada, Las Vegas, USA

REFERENCES

Calvo, R. A., D'Mello, S., Gratch, J., & Kappas, A. (Eds.), (2015). *The Oxford handbook of affective computing*. New York: Oxford University Press.

Ivonin, L., Chang, H.-M., Diaz, M., Catala, A., Chen, W., & Rauterberg, M. (2015). Beyond cognition and affect: Sensing the unconscious. *Behaviour & Information*, 34(3), 220–238.

Kim, J., & Ryu, H. (2014). A design thinking rationality framework: Framing and solving design problems in early concept generation. *Human-Computer Interaction*, 29(5–6), 516–553.

Norman, D. A. (1990). *The design of everyday things*. New York, NY: Doubleday.

Pacey, A. (1983). *The culture of technology*. Cambridge, MA: MIT Press.

Tettegah, S., & Espelage, D. (2015). (Forthcoming, December). *Emotions, behaviors and digital media: Communication of feelings through, with and for technology: Vol. 4*. London: Elsevier Publishers.

Tettegah, S., & Gartmeier, M. (2015). (Forthcoming, November). *Emotions, design, learning and digital media: Communication of feelings through, with and for technology: Vol. 2*. London: Elsevier Publishers.

Tettegah, S., & Huang, W. D. (2015). (Forthcoming, October). *Emotions, games and digital media: Communication of feelings through, with and for technology: Vol. 1*. London: Elsevier Publishers.

Winner, L. (1986). Do artifacts have politics? In *The whale and the reactor: A search for limits in an age of high technology*. Chicago, IL: University of Chicago Press (pp. 19–39).

SECTION I

Experiments and Theories in Emotions, Technology, and Design

CHAPTER 1

Emotional Screen: Color and Moving Images in Digital Media

Federico Pierotti
University of Florence, Firenze, Italy

From the very beginning of the medium of film, up to the end of the nineteenth century, the use of color in motion pictures has been supported by scientific and experimental knowledge on bodily and emotional responses to color stimuli. Today, this bond is strongly evident in various areas of film and media production. Indeed, contemporary cinema displays a predominant tendency deriving from the emotional impact of color in connection with the cultural dissemination of neuroscience and with the new color manipulation possibilities presented by the recent digital turn.

The particular objective of this essay is to highlight that there is a specific means of expressing color in contemporary cinema that may be defined with the concept of *emotional color*. The substantiation of such aesthetics was undoubtedly furthered by the recent digital turn, which not only stands as a technological breakthrough, but marks a real change in perception (Casetti, 2008, p. 187; Ritchin, 2009; Rodowick, 2007). Indeed, with the transition from analog to digital, color loses its indexical relationship with profilmic reality (Manovich, 2001, p. 300), and this aspect has reinforced the idea that color is to be perceived by the viewer in an emotional sense. Color has gained increasing autonomy, becoming a key element for the use of images in contemporary film and media.

Today's cultural interest in the subject of color has been fueled by the combination of several factors. Firstly, studies in neuroscience on the visual brain and the neural basis of perceptual processes have shown that the origin of receptive vision lies in color (Zeki, 1999); the spread of neuroscience across significant areas of contemporary society has encouraged the circulation of such knowledge in mass culture.

Secondly, contemporary cinema reveals a clear interest in episodes of film history where color was regarded as a sensory element (color as attraction in early cinema, the prelogical thinking of color in Ejzenštejn, "color of feelings" in Antonioni). Furthermore, postmodern aesthetics call us to consider perception as a bodily experience. The idea of it being the body that is perceiving[1] ties in with the idea that color is no longer a mere visual sensation of the eye, but a synesthetic experience engaging all the senses.

On the basis of the above assumptions, this essay will focus on some specific choices of color and black-and-white design through which the emotional impact of color will be measured in contemporary digital cinema and video. These choices are designed to evoke empathy or detachment in the viewer. In order to study the relationships between emotions and colors on the screen, reference should be made to the dialog between symbolic conventions and theories on the neurobiological responses to color stimuli.

Digital cinema and video will be considered as a useful tool to understand contemporary visual culture.[2] Studying color in this perspective means going beyond a purely formalist approach and gaining an understanding of two important aspects: the origin of cultural knowledge used by features and shorts, be it intentionally or less so, and the approach it adopts in relation to such knowledge.

COLOR AS CINEMATIC EMOTION: A HISTORICAL SUMMARY

The subject of the relationship between color and emotion is not exclusive to contemporary culture; rather, it has been closely tied to the nature of modern visual media since their respective origins. The media of photography and film expanded over a period from the mid- to late-nineteenth century, which saw significant epistemological changes relating to the relationship between technology and forms of perception (Casetti, 2008; Crary, 1990, 1999). Film is a new experience, characterized by its strong impact on the senses and the body of the viewer and, from the outset, color was used to increase the intensity of this impact (Gunning, 1994). In early cinema, the attraction effect of color is one of the sensory experiences through which theories, experiments, and ideas on color are applied and disseminated (Yumibe, 2012). Subsequently, the subject of the impact of color

---

[1] On this point, see Sobchack (2004), Elsaesser & Hagener (2010), Rocha Antunes (2015).
[2] Among the most important studies on contemporary digital media, see, above all, Bolter & Grusin (1999), Jenkins (2006). On the emotional impact of popular culture, see also Jenkins (2007).

on viewers continued to be of interest both to experts of science and technology (inventors, scientists, industrialists) and operators of the creative industry (directors, cinematographers, art directors, color consultants).

Between the 1930s and 1960s, there was a slow and gradual transition from black-and-white to color in primary modern visual media (film, print, television, photography). The various factors that brought about this shift may be seen in a new light when considered on the basis of contemporary knowledge on the psychological and behavioral significance of color. For instance, in the period where Hollywood's mode of representation became established, it was crucial to control the emotional intensity of the medium of film, making it second to the development of the narrative. The ideal form for this model is black-and-white, which predominated up until the fifties. When Hollywood classical cinema came into contact with Technicolor, between the 1920s and 1930s, the question of color constantly returned back to the relationship with the viewer. Empirical experience and scientific knowledge on the impact of color were used, with varying degrees of effectivity, in support of the contrasting viewpoints. Those who opposed Technicolor were convinced that color acted as a barrier when watching a film, as it attracts the viewers' attention away from the narrative. Whereas, its proponents argued that color was able to enrich the sensory and emotional experience of the viewer (Kalmus, 1935). In the context of this debate, the introduction of Technicolor into Hollywood had to undergo lengthy negotiations with the studio system, through which a model was established based on color restraint and controlling its emotional power (Higgins, 2007). Even with regard to other media, color has often been considered as an element that is to be introduced in small doses, given the widespread belief in its ability to distract. The colors in media which were more readily accepted in cultural terms were often those that aimed to imitate the balance of perception of human vision, which were essentially less noticeable than others.[3]

Alongside this predominantly cultural model, there was a prelogical and emotional undercurrent of color flowing through the history of cinema; this,

---

[3] Another issue to be considered is the way in which skin color was enhanced and compromised by the technology. As Richard Dyer (1997) has remarked, "The photographic media and, *a fortiori*, movie lighting assume privilege and construct whiteness. The apparatus was developed with white people in mind and habitual use and instruction continue in the same vein, so much so that photographing nonwhite people is typically construed as a problem. All technologies work within material parameters that cannot be wished away. Human skin does have different colors which reflect light differently" (p. 89).

nonetheless, remained marginal. It emerged from time to time in specific areas of production and consumption, such as advertising, or through individual filmmakers such as Antonioni, Demy, Ejzenštejn, and Godard.

In contemporary cinema and media, this emotional thread of color has taken on great importance, as it was in the days of early cinema (Yumibe, 2012, p. 151). Increasingly, in today's audio-visual products, the functions of color are not second to the logic of the narrative, nor do they seek to mimic the conventional patterns of human perception. A renewed desire to take full advantage of the sensory and emotional impact of color appears to be giving rise to a complete reversal of the previous balance: emotional color is designed and used as a sensory and bodily stimulus targeted at the viewer.

FROM BRAIN TO MARKETING: THE ATTRACTION EFFECT

Color is able to play a crucial role in both engaging attention and as an element to elicit emotion. At least since the time of Goethe, who had spoken of the moral action of color, a great deal of theoretical and experimental insights aimed to emphasize the close bond that seems to exist between color and attention and between color and emotion. Today, neurosciences and their applications in the area of social sciences are seeking neurobiological explanations for these connections.

Neurosciences have produced scientific evidence on the functional specialization of the areas of the brain. Of these, the findings on the visual brain and areas of the brain involved in perceiving color are particularly noteworthy (Bartels & Zeki, 2000; Lueck et al., 1989; McKeefry & Zeki, 1997; Viviani & Aymoz, 2001; Zeki et al., 1991). These studies have shown the sequencing of the processes, highlighting how color is the primary perceptual element in engaging brain stimulus. A similar discovery has confirmed something that, since the late nineteenth century, had often been expressed on the theoretical level or supported on the basis of empirical and experimental evidence: the primary function of color in capturing attention and enticing the senses. These neuroscientific studies initially focused on monodimensional perceptual experiences, thereby making it possible to isolate and study the various data of human perception separately. In recent years, experiments dealt with increasingly complex trends, highlighting the brain interactions that are engaged during the simultaneous perception of multiple stimuli. In this respect, reproductions of paintings or film sequences have been used to formulate hypotheses on brain functions when dealing with visual and sound constructions that are particularly rich in stimuli (Bartels & Zeki, 2004a, 2004b; Risko, Laidlaw, Freeth, Foulsham, & Kingstone, 2012).[4]

Part of this research focused on elements within the images that are able to activate brain stimulation of a given entity. Some studies have demonstrated that these processes are related to the appearance of new objects in the field of vision. Others have shown that, with the elements involved in this context, one must also consider the variations in formal factors, in particular those relating to light and color (Franconeri, Hollingworth, & Simons, 2005; Sàenz, Buračas, & Boynton, 2003; Snowden, 2002). For the purposes of this study, it is interesting to note that some particular forms of stimulus, which deviate from the stimulation deemed to be most commonplace or conventional, have been associated with particular modes of neural excitation. For example, some experiments have shown that the perception of colors that are considered unnatural or detached from typical perceptual experiences (such as blue strawberries) gives rise to a different brain stimulation than that which can be seen in the presence of objects deemed to be natural (red strawberries). Similar findings concentrate on the role played by expectations, memory, and past knowledge on perceptual and cognitive processes relating to color (Zeki & Marini, 1998).

David Katz (1935), in his gestalt studies on how color appears, had already studied the psychological phenomenon of color memory, which occurs when an object is shown to the senses in a color other than that in which it is usually seen (p. 160). It is interesting to understand, in this respect, how the world of images and media potentially has very large scope. Director Michelangelo Antonioni had realized its possibilities in 1964, in his first color film, *Red Desert* (*Il deserto rosso*). The film explores the ambiguous nature of the relationship between the way color appears, dealt with by Katz, and the bodily and emotional responses to color stimuli, at the center of contemporary thinking in the field of applied psychology. To break the link between the perceived color under natural conditions, and what Antonioni called "the color of feelings," the director had to repaint objects, architecture, and landscape features. The reflection on the emotional nature of color presented by this film has returned today through the current importance that color is taking on once again in digital media and cinema. The endless possibilities of manipulating and permuting color make it possible to break the expected link between an object and its color as it is normally perceived.

---

[4] For an overview on the impact of neurosciences on cognitive film theories and film studies, see also Gallese & Guerra (2012).

The theories developed in the area of neuroscience are also not currently exclusive to scientists and specialists, but rather are wide spread in social sciences and practices applied by the industry. One area where it is possible to witness a strong interest in knowledge on the impact of color on attentional, emotional, and behavioral processes is in marketing and advertising (Babin, Hardesty, & Suter, 2003; Bagchi & Cheema, 2013; Crowley, 1993). Once again, this is not a new or recent phenomenon, but a line of studies rooted in the very birth of modernity (Blaszczyk, 2012). The idea of using the color of an object or image as an element capable of attracting attention becomes paramount at the end of the nineteenth century, in the context of a visual culture marked by the establishment of distracted perception (Crary, 1999, p. 46).

This same idea becomes even more important after World War II, when the production of industrial objects in different color variants becomes generalized, even for those objects where color was initially considered to be a secondary element (e.g., cars). At the same time, color reproduction technologies make it increasingly easier and cheaper to obtain color images (Enticknap, 2005, p. 74), which, in turn, make it possible to use media to emphasize the role of color selection as an emotional factor. Just in the years when the major twentieth-century mass media began to convert to color, a good deal of studies, influenced by the psychology of perception, reflect on the potential effects and effectiveness of color stimuli in a very wide range of public and private social contexts, from the workplace, the home, to shops and hospitals (Birren, 1945, 1969).

In contemporary visual culture, the spread of digital images and the viewer's immersion in a world of stimuli in ever-increasing quantity make such issues even more relevant. For some years now, the scientific findings on the brain have given rise to a new line of studies called neuromarketing (Nelson, 2008, p. 123). This goes back over a great deal of issues—which had already undergone empirical testing in the past—in the light of neurosciences. Using this approach, an attempt is made to understand which are the stimuli induced by objects, images, sounds, lights, colors, and other elements found in sale areas or in advertising. These studies reflect on the ability of images, logos, objects, packaging and installations to produce what is known as the attraction effect. In a visual culture marked by an increasing number of stimuli and increasingly short attention spans, the ability to attract attention through a powerful and effective stimulus is becoming an even more important requirement than in the past. With attention fluctuating more and more between different media and devices, the risk of new distractions is always present. For this reason, it is becoming equally important to prevent decreasing attention and maintain the interest of the consumer and viewer. In this respect, together with the attraction effect, the ability to elicit positive emotions is particularly significant: some studies of neuromarketing tend to confirm previous findings on the role of emotions as a factor that significantly affects the behavior of potential consumers (Hedgcock & Rao, 2009).

PATTERNS OF EMOTION: THREE EXAMPLES FROM CONTEMPORARY DIGITAL MEDIA

Entering into the merit of these theories does not fall within the objectives of this essay, or within the competency of the author. What is of interest for film and media studies is highlighting how the current success and good fortune of this scientific knowledge applied to the social sphere may be seen through their dissemination in visual culture and mass culture. In this regard, visual media provide a potentially very fruitful and thought-provoking area for research (Brown, 2013). Consider, for example, the triumph of television products such as *Brain Games* (National Geographic Channel), or the relatively recent success of the so-called mind game films (*Fight Club*, David Fincher, 1999; *Memento*, Christopher Nolan, 2000; *Artificial Intelligence: AI*, Stephen Spielberg, 2001; *Eternal Sunshine of the Spotless Mind*, Michael Gondry, 2004, see Elsaesser, 2009). Media products such as these are paramount in order to study the spread of scientific theories within a wide array of knowledge and practices conferred to cultures and common feeling.

For the creators of moving images, the question of color's sensory impact becomes a practical problem that each time is to be addressed through specific color design choices. Directors, cinematographers, and colorists can be considered to be real problem-solvers, being called upon to develop creative solutions to ensure that their products achieve the appropriate emotional shade. Every media or film genre may be correlated with a series of basic emotions, each in varying proportions (Carroll, 1999; Grodal, 2009; Tan & Frijda, 1999): a comedy will work primarily on happiness, a revenge movie on anger, a melodrama on sadness, a horror or war movie on fear, a thriller on surprise, and a gore film on disgust. In this regard, color design choices interact with other elements of the mise-en-scène (shots, camera movements, editing, sound, and music[5]) with the aim to evoke in the viewer

---

[5] On the impact of sound and music in contemporary video and cinema, see Vernallis (2013).

the reactions of empathy or detachment that are deemed most appropriate in each case. These choices are influenced by what we know, or what we think we know, about the emotional impact of color. It is very often the case, in mainstream production, that each set of emotions is correlated with stereotyped and culturally shared color associations: livid colors for horror films and thrillers (*The Ring*, Gore Verbinski, 2002; *The Village*, M. Night Shyamalan, 2004), desaturated colors for the war movies (*Saving Private Ryan*, Steven Spielberg, 1998; *Flags of Our Fathers*, and *Letters from Iwo Jima*, Clint Eastwood, 2006), warm colors for comedies (*Amelie*, Jean-Pierre Jeunet, 2001), and saturated colors for melodramas (*Far from Heaven*, Todd Haynes, 2002). The digitalization of cinema (Harbord, 2002, p. 138) makes these choices of color and design increasingly refined, and at the same time, increasingly important. Over the past 15 years, digital color correction was canonized as a creative procedure that is able to redesign the color of an entire movie (Belton, 2008; Higgins, 2003; Misek, 2010, p. 152). The most famous colorists have taken on a leading role in constructing the visual aspect. See, for example, cases such as *O Brother Where Art Thou?* (Joel Coen, 2000) and *The Aviator* (Martin Scorsese, 2004), where the color look has been completely redesigned thanks to digital technology, to simulate, in the former, the appearance of an old colored postcard (Brown, 2013, p. 215; Fisher, 2000) and, in the latter, the look of the Technicolor films of the twenties and thirties (Higgins, 2007, p. 217; Pavlus, 2005).

One area of visual culture where color design choices have taken on a very important role is in media advertising. Yet, the interest in this aspect is not new. Since the birth of chromolithography in the nineteenth century, the field of advertising has always been engaged in experimenting with new color reproduction technology. A similar phenomenon has focused on the digitalization of visual media, which began in the 1990s (Enticknap, 2005, p. 202): the advertising short was one of the first media formats to convert to digital (Misek, 2010, p. 159) and take advantage of the full potential of the new technology in terms of empathy and emotional impact on the viewer. The advertising industry is, therefore, one of the key areas for understanding the features of contemporary emotional color.

For this reason, I have chosen to analyze two recent advertising shorts with differing color design choices: *Gucci Bright Diamante*, in color (2013), and *Renault Clio Costume National*, in black-and-white (2013). These two examples clearly highlight the features of advertising shorts designed for interactive media devices, such as computer displays, tablets, or mobile phone: short and fast communication, audiovisual stimuli, quick and continuous pace, presence of music and drastic limitation of verbal elements. The third example relates to the mixture of the two forms, color and black-and-white, within the same image, in several sequences of the film *Sin City* (Frank Miller and Robert Rodriguez, 2005). These three examples represent three signs through which one can gain an understanding of the current importance of the emotional aspect of color. Despite their differences, what the three examples have in common is the intention to create unconventional color designs, through which the viewer is invited to have real sensory experiences, in a separate perspective from his or her perception of the natural world.

The first example is the advertising short *Gucci Bright Diamante*.[6] The first shot is made up of split screen of nine different images, suggesting the idea of a three-story house. At the center of the bottom row, we see the image of a white door on a white wall, while the other eight images show the same number of rooms with monochrome décor and walls, as well as male and female models posing with clothes of the same color. The monochrome is broken by the bags and suitcases, disorderly arranged in the various rooms. A first series of shots in rapid succession shows the models as we had seen them in the initial split screen. Subsequently, the models begin to move from one room to another, creating new color combinations. The last set of shots showing the models once again in their initial rooms, and the handbags and suitcases are finally in order. The final shot repeats the split screen: each bag is now in the room corresponding to its color.

The construction of the commercial may be reduced to four basic formal elements: color, music, movement, and editing. The presence of saturated and spot colors, continuous movement, quick editing and music help to induce a multisensory perception, based on the synesthetic interplay between vision, hearing, and touch. The division of the screen into several windows, in the first and final frame, repeats the typical spatial set-up of interactive displays, such as touch screens, which use the haptic power of color (Lundemo, 2006) to identify the various functions of the device. In this case, it seems to engage a form of interactive stimulation similar to that of certain video games: the purpose of the game is to create the monochrome in each room, moving each bag or suitcase into the correct room. The viewers do not have the ability to actually play the game, that is to say, touch

---

[6] www.gucci.com/uk/worldofgucci/videos/bright-diamante (last access: 13/01/2015). It has a duration of 65 s and consists of approximately 55 shots, which follow the rhythm of insistent and hypnotic techno music. The logo appears at the opening and the end.

and move the bags, but the models can do it for them. The short repeatedly plays on stimulating the sense of touch. Several shots show the models grabbing or exchanging the bags, thereby producing a process of identification: thanks to the interaction between these actions and the haptic use of color, it is as if the viewer were invited to touch the bags and indirectly experience the thrill of possessing them. Furthermore, the fact that the number of female models is greater than that of male models clearly indicates that the target audience of the advertisement is predominantly female.

The second example relates to the advertising short *Renault Clio Costume National*.[7] The narrative is much more distinct and is supported by linear editing; the two main characters can be identified from the first shot: a man and a woman. The pair exchange glances, she abandons her bags and gets into a taxi; he takes the bag and starts driving his car through a modern urban landscape. We see the woman again in a meeting room, where the man arrives showing the bag. The man, smiling, is then once again at the wheel of his car with the woman sitting next to him, laughing.

The narrative brings back two stereotypes of classic cinema: pursuit and seduction. The sequence of shots is primarily made up of eyeline matches, building a narrative and sequential connection between the various moments. The sense that is being engaged is, above all, sight. The sense of touch is engaged only twice: when the man's hand grabs the bag left behind by the woman, and when he selects reverse gear. While the previously analyzed bags short emphasized the haptic and tactile power of color, this seems to focus on the ability of black-and-white to create an aesthetic and contemplative detachment. The first had spot colors and sharp contrasts, whereas this color design is based on more subtle contrasts on the gray scale. Through the reflections of the buildings on the black surface of the car and transparent windows, abstract visual patterns are created that are constantly changing. The car becomes an aesthetic object to be contemplated, a work of art in motion and, at the same time, the main object toward which the viewer's visual attention is channeled. The black color of the car adds another positive connotation, as it evokes a series of positive cultural values, which are traditionally attributed to black in Western culture: elegance, simplicity, and good taste. The short is indicative of the role played by black-and-white in contemporary visual culture. Indeed, in the digital world,

---

[7] www.renault.it/cliocostumenational/ (last access 13.01.2015). It has a duration of 51 s and is made up of 32 shots in black and white, plus the brand appearing at the end, in color on a white background. The accompanying music is the song *The Guesser* by The Temples.

black-and-white is no longer a piece of specific and stand-alone technology, rather it has become one of the possible functions of the color image; as we know, each image is made up of pixels, which in turn contain numerical encodings relating to the three primary colors. This technological aspect has not removed black-and-white from our visual landscape; quite the reverse, it seems to have reinforced its cultural importance. In recent years, many media products have often used black-and-white.

The third case that is particularly noteworthy is where black-and-white is mixed with color within the same image (Schmerheim, 2013). This sort of border between color and black-and-white life, albeit using far more advanced technology, revives the same perceptual effect of hand and stencil coloring practiced in early cinema (Manovich, 2001, p. 296). The revival of this method began in the nineties, when digital technologies began to take on a central role in the postproduction phase. In the first pioneering examples, the appearance of color can still be read in emotional terms. In the famous sequence of *Schindler's List* (Steven Spielberg, 1993), the little girl's red coat shows a silhouette of color within the black-and-white shots. These shots are connected with Schindler's gaze: the color indicates that the coat attracts the attention of the protagonist and produces an intense emotional shock in his mind. Whilst *Schindler's List* is, nevertheless, still a film in black-and-white, where color has merely a limited appearance, at the end of the nineties, the mix of color and black-and-white was used to guide the narrative structure of an entire film: *Pleasantville* (Gary Ross, 1998). In this case, the link between color and emotion is made explicit by the unexpected eruption of color in the world of the American television series of the fifties: while black-and-white is associated with the moralistic and repressive world devoid of affection, the progressive burst of color in the city indicates the flourishing of erotic and emotional energy of its inhabitants. The film was shot entirely on color film and was chromatically realigned during the digital postproduction: all sequences relating to the television series, which were to appear in black-and-white, were discolored, while those that indicated the gradual irruption of colored objects and bodies in the world of *Pleasantville* were selectively discolored (Fisher, 1998, p. 62; Kaufman, 1999, p. 128).

After these pioneering experimental cases, mixing color with black-and-white became a color design technique that is relatively simple to achieve. This has been used more and more frequently in contemporary media production, especially in advertising (see, for example, the recent advertising campaign by T-Mobile). A particularly interesting example in cinema is

Sin City (Schmerheim, 2013, p. 121). This film uses clear, well-defined black-and-white with no shading. This visual choice refers to the stylized language of the graphic novel of the same name, on which the film is based, and helps to create an artificial world permeated by impulses as well as basic and violent passions; good against evil, and beautiful against ugly. Similarly, bright and primary colors alternate to highlight the highly intensive emotions and shocks experienced by the characters. Many sequences in the film are, in fact, punctuated by the presence of digitally colored objects and details in primary colors: red, yellow, blue, and green.

In the first sequence, we see a young woman looking over the railing of a large terrace, which faces the spectacular skyline of a city. Amid the whites and blacks, the attention is directed toward the girl with her flame red dress and lipstick. The scene continues with the arrival of a young man who begins talking to her; when the flame of the lighter approaches the woman's face to light the cigarette, her eyes become colored in green. Soon after, when the two kiss, a frontal shot transforms them into backlit white silhouettes; from the black background of the city, the illuminated windows of the skyscrapers emerge as small white rectangles. A gunshot suddenly puts an end to this romantic idyll.

The sequence plays on the cultural stereotypes of the two colors that appear (the complementary red and green, in addition to white and black) to emphasize the ambivalence of the situation being shown. Red is associated with the female body and connotes passion and excitement, while at the end of the sequence, the meaning is reversed, evoking a sense of blood and violence. Similarly, the green of her eyes initially seems to allude to hope (of seducing a woman and being protected by the man). Yet, if read in the light of the final murder, this is overturned to connote negative meanings of adverse fate and money. The man was an assassin paid to kill the girl; he did not know her identity. Many other sequences in the film have a similar use of a given set colors that constantly appear in a flat and dematerialized form. These colors draw attention to the movement of objects (a pair shoes, a car, a suit), faces (eyes, hair), and bodies (the yellow complexion of the rapist). Their function, however, is not confined merely to visual stimulation, but often requires a connection with the sensory channels of hearing and smell. A somewhat epidermal stimulation is invoked whenever bodies torn by bullets or knives release splashes of red-colored blood. As in the opening sequence, the whole movie plays on the power of attraction of red and its strong symbolic ambivalence: while red is able to arouse empathy when it is associated with the sphere of desire and passion, it often causes reactions of disgust and revulsion when it is related to bloody and brutal acts. The sense of smell is, on the other hand, linked to the yellow complexion of the rapist. In one sequence in particular, following a shooting, the character leaves organic yellow traces considered nauseating by Bruce Willis' character. Even in this case, this grisly synesthetic interplay is held up by a series of negative cultural connotations; for centuries in Western culture, yellow has been considered the color of sickness and decay.

CONCLUDING REMARKS

Thanks to digital color correction, color has become an element which may be manipulated in a way that was previously unthinkable. The advertising industry has taken full advantage of this possibility, and this has become a central area for technical and expressive experiments, which for several years now, have become commonly used in film and other visual media.

In light of this change, this contribution has sought to observe that the subject of color in audiovisual products, such as films and advertising shorts, may be reconsidered in the light of a new integration between aesthetic, scientific, and technological disciplines. Indeed, the findings of neuroscience take on a new and potentially very fruitful approach to the question of the emotional impact of color on the viewer. It has, therefore, become important to question to what extent the cultural success of neuroscience currently helps to rethink the wide array of theories, knowledge, and practices on the relationship between color and emotion. The spread of this knowledge provides new knowledge and cognitive schemata on color which contemporary technologies are constantly required to address.

Consequently, it is equally important to study the forms and processes through which visual media are able to creatively construct this same knowledge. These phenomena are, in fact, shaping a new kind of sensitivity toward color. In this essay, I have sought to examine some of the potential relationships found between the area of science and topics on color and emotion and the tendency of digital cinema to develop color models with a strong visual and sensory impact.

Many contemporary media products (films, television series, images, and advertising videos, video clips) show a strong interest in the emotional impact of color, which, to a marginal extent, is emerging as a dominant trend, as evidenced by the sequences analyzed. This trend implies a model where the cognitive activity of the viewer is focused on engaging his or her entire sensory apparatus and on the centrality of emotional responses.

As such, the cultural structure of emotional color is symptomatic of the spread of contemporary knowledge on the relationship between color and emotion. As the analyses put forward have helped to demonstrate, the most important aspect is the extreme intensification of all those elements that combine to engage attention and produce a high degree of emotional arousal and multisensory involvement. The three examples chosen have shown three possible ways of building artificial and self-referential color patterns, which are detached from the standard conditions of conventional vision: saturated, primary, and spot colors, recalling the abstract and immaterial palette of digital technology (*Gucci Bright Diamante*); black-and-white playing with reflections and changing transparencies (*Renault Clio Costume National*); and the mixture of a limited range of saturated colors in graphic black-and-white with no shading (*Sin City*).

REFERENCES

Babin, B. J., Hardesty, D. M., & Suter, T. A. (2003). Color and shopping intentions: The intervening effect of price fairness and perceived affect. *Journal of Business Research*, *56*(7), 541–551.

Bagchi, R., & Cheema, A. (2013). The effect of red background color on willingness-to-pay: The moderating role of selling. *Journal of Consumer Research*, *39*(5), 947–960.

Bartels, A., & Zeki, S. (2000). The architecture of the colour centre in the human visual brain: New results and a review. *European Journal of Neuroscience*, *12*(1), 172–193.

Bartels, A., & Zeki, S. (2004a). Functional brain mapping during free viewing of natural scenes. *Human Brain Mapping*, *21*(2), 75–85.

Bartels, A., & Zeki, S. (2004b). The chronoarchitecture of the human brain: Natural viewing conditions reveal a time-based anatomy of the brain. *NeuroImage*, *22*(1), 419–433.

Belton, J. (2008). Painting by the numbers: The digital intermediate. *Film Quarterly*, *61*(3), 58–65.

Birren, F. (1945). *Selling with color.* New York: McGraw-Hill.

Birren, F. (1969). *Light, color and environment: A thorough presentation of facts on the biological and psychological effects of color.* New York: Van Nostrand Reinhold.

Blaszczyk, R. L. (2012). *The color revolution.* Cambridge, Massachusetts: MIT Press.

Bolter, J. D., & Grusin, R. (1999). *Remediation: Understanding new media.* Cambridge, Massachusetts: MIT Press.

Brown, W. (2013). "Those men are not white!": Neuroscience, digital imagery and color in O brother, where art thou? In S. Brown, S. Street, & L. Watkins (Eds.), *Color and the moving image: History, theory, aesthetics, archive* (pp. 209–218). New York: Routledge.

Carroll, N. (1999). Film, emotion, and genre. In C. Plantinga & G. M. Smith (Eds.), *Passionate views: Film, cognition, and emotion* (pp. 21–47). Baltimore: Johns Hopkins University Press.

Casetti, F. (2008). *Eye of the century: Film, experience, modernity.* E. Larkin & J. Pranolo, Trans New York: Columbia University Press.

Crary, J. (1990). *Techniques of the observer: On vision and modernity in the nineteenth century.* Cambridge, Massachusetts: MIT Press.

Crary, J. (1999). *Suspensions of perception: Attention, spectacle, and modern culture.* Cambridge, Massachusetts: MIT Press.

Crowley, A. E. (1993). The two-dimensional impact of color on shopping. *Marketing Letters*, 4(1), 59–69.
Dyer, R. (1997). *White*. New York: Routledge.
Elsaesser, T. (2009). The mind-game film. In W. Buckland (Ed.), *Puzzle films: Complex storytelling in contemporary cinema* (pp. 13–41). Chichester, West Sussex; Malden, MA: Wiley-Blackwell.
Elsaesser, T., & Hagener, M. (2010). *Film theory: An introduction through the senses*. New York: Routledge.
Enticknap, L. (2005). *Moving image technology: From zoetrope to digital*. London: Wallflower Press.
Fisher, B. (1998). Black-and-white in color. *American Cinematographer*, 79(11), 60–67.
Fisher, B. (2000). Escaping from chains. *American Cinematographer*, 81(10), 36–49.
Franconeri, S. L., Hollingworth, A., & Simons, D. J. (2005). Do new objects capture attention? *Psychological Science*, 16(4), 275–281.
Gallese, V., & Guerra, M. (2012). Embodying movies: Embodied simulation and film studies. *Cinema: Journal of Philosophy and the Moving Image*, 3, 183–210.
Grodal, T. K. (2009). *Embodied visions: Evolution, emotion, culture, and film*. Oxford: Oxford University Press.
Gunning, T. (1994). Colorful metaphors: The attraction of color in early silent cinema. *Fotogenia*, 1(1), 249–255.
Harbord, J. (2002). *Film cultures*. London, Thousand Oaks: Sage Publications.
Hedgcock, W., & Rao, A. R. (2009). Trade-off aversion as an explanation for the attraction effect: A functional magnetic resonance imaging study. *Journal of Marketing Research*, 46, 1–13.
Higgins, S. (2003). A new color consciousness. *Convergence*, 9(4), 60–76.
Higgins, S. (2007). *Harnessing the Technicolor rainbow: Color design in the 1930s*. Austin: University of Texas Press.
Jenkins, H. (2006). *Convergence culture: Where old and new media collide*. New York: New York University Press.
Jenkins, H. (2007). *The wow climax: Tracing the emotional impact of popular culture*. New York: New York University Press.
Kalmus, N. (1935). Color consciousness. *Journal of the Society of Motion Pictures Engineers*, 25 (2), 139–147.
Katz, D. (1935). *The world of colour*. R.B. MacLeod & C.W. Fox, Trans London: Kegan Paul.
Kaufman, D. (1999). Creating a digital film lab. *American Cinematographer*, 80(9), 128–129.
Lueck, C. J., Zeki, S., Friston, K. J., Deiber, M. P., Cope, P., Cunningham, V. J., et al. (1989). The colour centre in the cerebral cortex of man. *Nature*, 340(6232), 386–389.
Lundemo, T. (2006). The colors of haptic space: Black, blue and white in moving images. In A. Dalle Vacche & B. Price (Eds.), *Color: The film reader* (pp. 88–101). New York: Routledge.
Manovich, L. (2001). *The language of new media*. Cambridge, Massachusetts: MIT Press.
McKeefry, D., & Zeki, S. (1997). The position and topography of the human colour centre as revealed by functional magnetic resonance imaging. *Brain*, 120(12), 2229–2242.
Misek, R. (2010). *Chromatic cinema: A history of screen color*. Chichester, U.K.; Malden, MA: Wiley-Blackwell.
Nelson, M. R. (2008). The hidden persuaders: Then and now. *Journal of Advertising*, 37(1), 113–126.
Pavlus, J. (2005). High life. *American Cinematographer*, 86(1), 38–53.
Risko, E. F., Laidlaw, K. E. W., Freeth, M., Foulsham, T., & Kingstone, A. (2012). Social attention with real versus reel stimuli: Toward an empirical approach to concerns about ecological validity. *Frontiers in Human Neuroscience*, 6(143), 1–11.
Ritchin, F. (2009). *After photography*. New York: W.W. Norton.

Rocha Antunes, L. (2015). Neural correlates of the multisensory film experience. In M. Grabowski (Ed.), *Neuroscience and media: New understandings and representations* (pp. 46–61). New York: Routledge.

Rodowick, D. N. (2007). *The virtual life of film*. Cambridge, Massachusetts: Harvard University Press.

Sàenz, M., Buračas, G. T., & Boynton, G. M. (2003). Global feature-based attention for motion and color. *Vision Research, 43*(6), 629–637.

Schmerheim, P. (2013). From Psycho to Pleasantville: The role of color and black-and-white imagery for film experience. In S. Brown, S. Street, & L. Watkins (Eds.), *Color and the moving image: History, theory, aesthetics, archive* (pp. 114–123). New York: Routledge.

Snowden, R. J. (2002). Visual attention to color: Parvocellular guidance of attentional resources? *Psychological Science, 13*(2), 180–184.

Sobchack, V. (2004). *Carnal thoughts: Embodiment and moving image culture*. Berkeley: University of California Press.

Tan, S. H., & Frijda, N. H. (1999). Sentiment in film viewing. In C. Plantinga & G. M. Smith (Eds.), *Passionate views: Film, cognition, and emotion* (pp. 48–64). Baltimore: Johns Hopkins University Press.

Vernallis, C. (2013). *Unruly media: YouTube, music video, and the new digital cinema*. Oxford: Oxford University Press.

Viviani, P., & Aymoz, C. (2001). Colour, form, and movement are not perceived simultaneously. *Vision Research, 41*(22), 2909–2918.

Yumibe, J. (2012). *Moving color: Early film, mass culture, modernism*. New Brunswick, N.J.: Rutgers University Press.

Zeki, S. (1999). *Inner vision: An exploration of art and the brain*. Oxford-New York: Oxford University Press.

Zeki, S., & Marini, L. (1998). Three cortical stages of colour processing in the human brain. *Brain, 121*(9), 1669–1685.

Zeki, S., Watson, J. D. G., Lueck, C. J., Friston, K. J., Kennard, C., & Frackowiak, R. S. J. (1991). A direct demonstration of functional specialization in human visual cortex. *Journal of Neuroscience, 11*(3), 641–649.

CHAPTER 2

Safe and Sound: Using Audio to Communicate Comfort, Safety, and Familiarity in Digital Media

Michael L. Austin
Howard University, Washington, DC, USA

As I constantly searched for a quiet place in which I could sit to write this essay, one fact became increasingly clear: sounds are *everywhere*. And understandably so, especially since sounds perform an important function in how we receive information from our environment—even expressive, emotional information. Sound has been used for centuries as both a way to express emotions and as a method to elicit emotional responses from listeners; from audible speech and visceral utterances, to expressive music in concerts, film, and theater, sound has always played a major role in affective interactions. Many everyday objects are constructed to make sounds with affective properties and are increasingly designed and utilized to communicate or elicit positive emotional responses in consumers of digital media. This is particularly true for objects that provide a sense of comfort or safety, as they signal successful user experience (UX) design.

For example, imagine stepping out of your car or a taxi and closing the door behind you as you hurry off to a meeting. Your visual imagination can probably conjure up many details regarding the high level of design that went into the manufacturing of your car, especially if it is a luxury car, such as BMW: sleek, aerodynamic contours, an idiosyncratic shade of black with metal flecks in the paint, fashionable leather seats, and a specially designed control interface with icons and fonts exclusive to BMWs. Just as design features convey a sense of style and luxury, the snug feeling of the seatbelts, the quick response of the breaks, and the digital screen that allows the driver to see behind the car are all designed to express a sense of safety. Within your aural imagination, can you hear the sound of the car door slamming shut? If this car is, in fact, a BMW, there is a good chance that the sound the door makes as it closes was designed by Emar Vegt, an "aural designer" who works at the company's head office in Munich. Describing this design element in an interview with David Baker for *Wired Magazine*, Vegt says, "The sound of the door closing is a remarkable aspect of the buying decision … It gives people reassurance if the door feels solid and safe" (Baker, 2013). Likewise, other sounds in BMWs are designed to be a little discomfiting as a reminder to think about safety or to account for the safety of others. Baker goes on to write:

> *Inside the car are other considerations. "Warning sounds need a particular aesthetic," he says [citing Vegt]. The noise that tells the driver to put on their seat belt can't be too pleasant as "people will listen to it like a symphony." But neither can they be too annoying—people find ways of shutting them off. And electric cars are a challenge. "Sounds tells [sic] people that a car is there, which is really important for blind people."*[1]
>
> *(Baker, 2013)*

In order to assure their car sounds the way it should, "every sound made by a BMW is analyzed by a team of over 200 acoustic engineers to ensure they are both mechanically and acoustically correct" (Jackson, 2003, p. 106).

Just as the closing of a car door is carefully designed to give the driver the feeling it has been safely shut and adds to a sense of confidence in the entire car, designers of audio for digital media often want to educe feelings of security and familiarity as this signals successful UX design. In this essay, I explore significant uses of sound and audio feedback to communicate feelings of comfort, security, and intimacy in digital media. Designers—sound designers, UX designers, human-computer interaction (HCI) designers, interface designers, and designers and engineers from a host of disciplines—often seek to create positive, engaging experiences with technology, and do so through the utilization of sonic material to ensure that a user's encounter with technology is a pleasant one. Using case studies involving interfaces designed to communicate safety and comfort, I frame listening, particularly semantic modes of listening, as the primary way in which the emotional information conveyed through digital media is understood,

---

[1] Here, Vegt is making reference to the fact that electric cars were found to be more dangerous to pedestrians than gas-powered cars because they were much quieter, almost silent, and the visually impaired were sometimes unaware of their presence. To address this problem, the European Parliament ruled that electric and hybrid cars must add artificial noise to their engines. See Walker, A. "Silent But Deadly: The EU Wants Electric Cars to Add Sounds for Safety." GIzmodo, April 7, 2014. http://gizmodo.com/silent-but-deadly-the-eu-wants-electric-cars-to-add-so-1560215281.

and I present rationale for considering listening, hearing, and the sounds themselves as equally necessary parts for an emotional understanding of digital media.

Before digging deeper into more examples of the possible ways in which this can be accomplished in digital media, primarily through auditory display, I first want to examine auditory awareness (i.e., the ways in which we are attuned to our sonic environments) in order to help frame an understanding of the ways in which one might listen to sounds from digital media within our environment, and how one might derive emotional meaning from them.

HEARING, LISTENING, FEELING

Aural information gathering, especially emotional information, does not simply depend upon a listener's ability to perceive sound as it propagates throughout an environment and process it physiologically. Rather, the ability to hear a sound physically, process it semantically, and understand it semiotically necessitates a broader understanding of the interconnected relationships of sounds, the way(s) we listen to them, and the environment from which they emanate. In *Background Noise: Perspectives on Sound Art*, Brandon LaBelle writes:

> Sound is intrinsically and unignorably relational: it emanates, propagates, communicates, vibrates, and agitates; it leaves a body and enters others; it binds and unhinges, harmonizes and traumatizes; it sends the body moving the mind dreaming, the air oscillating. It seemingly eludes definition, while having a profound effect.
>
> *(LaBelle, 2008, p. ix)*

We use sound to connect with others emotionally. Even before we are born, we are able to hear our mother's voice, distinguish it from other voices, and react to it (Kisilevsky et al., 2003). Lullabies are sung to infants and young children to help them feel connected to the singer and to give them a sense of safety and comfort. Studies indicate that music can help children relax, diminish their pain, and reduce their anxiety (Longhi, Pickett, & Hargreaves, 2015).

Clearly, sound has an impact on how we feel, but how are vibrations of air molecules able to bring us to tears, strike terror in our hearts, or make us feel safe? Throughout our lives, sound facilitates the expression of emotional communication with others; our ability to perceive any sonic information relies on our ability to process the auditory information we collect from

our environment. Researchers in human hearing, psychoacoustics, and cognition have identified several major skills that comprise healthy auditory processing. According to the American Speech-Language-Hearing Association, these skills are:
- Sound localization and lateralization (identifying the *place* of sound).
- Auditory discrimination (identifying different sounds and the differences among them).
- Auditory pattern recognition (identifying patterns in iterations of sounds, including those identified within Gestalt theory, such as grouping, figure/ground, good continuation, expectancy, etc.).
- Temporal aspects of audition, including:
  * temporal resolution (detecting a rapid succession of consecutive sounds as separate, rather than a single sonic event),
  * temporal masking (a process wherein sudden changes in the volume of a sound can "mask," or hide, the sonic event preceding or following it from our hearing),
  * temporal integration (combining patterns of sounds, or recognizing the contours that comprise a sound's envelope—attack, decay, sustain, or release—and translating that into useful information), and
  * temporal ordering (distinguishing the order in which successive sounds are heard).
- Auditory performance in competing acoustic signals (including dichotic listening; the ability to focus attention on important sounds and ignore background noise)[2] and
- Auditory performance with degraded acoustic signals.

(American Speech-Language Hearing Association, 1996, *parenthetical definitions mine*).

While auditory processing is a multimodal set of purely physiological activities, understanding the emotional content in sonic information, particularly in speech, relies heavily on these processes, especially since many of the ways in which emotions are communicated is through sound, such as tone of voice, sighs, tempo and volume of speech, vocal cadence, etc. Further, not only does central auditory processing disorder cause those with it to struggle with one or several of the skills listed above, resulting in difficulty understanding spoken communication, it can also possibly lead to a misunderstanding or misreading of the emotional cues embedded within speech.

---

[2] Also known as the "cocktail party effect."

The large and growing body of literature in psychological research on the affective/emotional qualities of music could also have implications for this examination of the emotional impact of sounds (both musical and otherwise) in digital media. Research in this area has been conducted for well over a century; recently, the approaches taken have either been to investigate possible causes of emotional arousal through music, or to investigate ways in which music might mediate the experience of emotions. Among causal factors, scholars list the listener's age, sex, music education, the physical environment wherein music is experienced, and whether or not the listener was alone or a member of an audience (Abeles & Chung, 1996; Gabrielsson, 2001), and some even question if there is something intrinsic within the music itself that elicits emotions (Sloboda, 1991). Others describe the underlying mechanisms that induce emotions, such as cognitive appraisal (Scherer, 1999), musical expectancy—(i.e., whether or not the music confirmed or defied listener's expectations (Meyer, 1956)), mental images that music could possibly evoke (Lyman & Waters, 1989; Osborne, 1980; Plutchik, 1984), and even brain stem reflexes and episodic memory (Juslin & Västfjäll, 2008). Although these and other approaches have not yet resulted in a single satisfactory conclusion regarding the source of music's emotional power, perhaps it accounts for some of the ways in which the same musical work can affect a wide array of seemingly contradictory emotions. As with auditory processing, these psychological approaches relate to the ways in which our ears perceive physical sound (i.e., how we actually *hear*), and possible ways in which our brains can perceive aural information and interpret it in an emotional manner.

Listening can, of course, be approached from more semiotic, philosophical, political, and theoretical perspectives as well, especially when considering our mutable relationship with sound. Because sound plays an important role in the way we experience the world, it also plays an equally vital role in the way we approach and begin to understand it. The most obvious connection between the philosophical and the ecological is made in the work of ethnomusicologist Steven Feld; he coined the term *acoustemology* (or acoustic epistemology), defined as "local conditions of acoustic sensation, knowledge, and imagination embodied in the culturally particular sense of place" (Feld, 1996, p. 91). In other words, one can understand the surrounding world, epistemologically, through its sounds. Pointing toward similar conclusions, Jacques Attali began his now-famous treatise, *Noise: The Political Economy of Music*, writing:

> For twenty-five centuries, Western knowledge has tried to look upon the world. It has failed to understand that the world is not for the beholding. It is for the hearing. It is not legible, but audible. Our science has always desired to monitor, measure, abstract, and castrate meaning, forgetting that life is full of noise and death alone is silent ... Now we must learn to judge a society more by its sounds ...
> **(Attali, 1985, p. 3)**

We can evaluate a culture and its values by examining its acoustic culture. In examining the sounds produced by our society, we begin to assign value to what we hear, most notably in the way we categorize the aural: silence versus sound versus noise, good versus bad, music versus Muzak, and eventually more emotional dichotomies such as pleasure versus annoyance, or even happy versus sad. Describing the ways in which categorizing the aural is a facet of social life, Ian Biddle writes:

> As a system by which the conceptual territories noise/music/silence are mapped and managed, the political ontology of sound is also a political theory of relationships: there is no quiet without less quiet, no noisy without less noisy, no music without its forbidden others. Class, ideology, race and gender are all visitors to this process of naming, of holding apart, and holding in mutually exclusive relation the three territories. They all make their way, like a little tiny parasitic relation of their own, into the mechanisms by which noise-obsessed neighbors, anxious public license granters, social theorists and policy makers seek to discipline and silence the social.
> **(Biddle, par 2)**

These attributes of acoustic culture correspond to the causal factors in previously mentioned research in music psychology, such as the way in which a listener's environment can possibly affect the mood or emotions that are elicited by music experienced in a particular time and place. Not only are these location-based causal factors used to contextualize musical and nonmusical sound, but they also present an ecological frame of reference for the emotional arousal triggered by them.

Thus far, I have described just a few of the myriad ways in which scholars, theorists, and researchers have attempted to label specific ways in which we attend to sound and account for emotional responses that result from this hearing. The sum of these parts that constitute the multimodality of hearing is listening, that is, *hearing to understand*. And just as there are many modes of hearing, listening, too, comes in many varieties. Pierre Schaeffer, a composer and one of the pioneers of the *musique concrète*[3] tradition, developed four

---

[3] Concrete music; this is a compositional practice wherein composers took both electronically-produced and found sounds, recorded them, and then manipulated them in an attempt to make them *acousmatic* (simply, sound that is heard but the listener cannot immediately identify its source).

modes of listening, or *Quatre Écoutes,* to illustrate the levels at which hearing, listening, and understanding are interrelated (Schaeffer, 1977). These levels are:
- *Ouïr:* This mode simply describes passive hearing. Sound waves strike our ears and we hear them, yet sometimes we do not seek to comprehend them. This applies to background noise and other environmental sounds that do not demand attention.
- *Écouter:* This is the most basic form of listening wherein we hear a sound and pay attention to it. Here, we consider sound in more semiological terms (What is the source of this sound? What does it indicate?).
- *Entendre:* This mode implies listening with intent; it is a combination of *ouïr* (an objective, physical hearing) and écouter (an objective, semiotic hearing); herein, a listener subjectively chooses the sound to which he or she pays attention.
- *Comprendre:* In this subjective mode, the listener prioritizes sound, decides which sounds are significant or irrelevant, and assigns meaning to sound. Here the sound object's essence is rendered irrelevant (i.e., What is special about this particular sound?), and that which is represented by the sound comes to the forefront (i.e., What does the sound *mean*? What does it *represent*?).

These four modes of listening are famously summarized by Schaeffer in one sentence: "I heard (ouïr) you despite myself, although I did not listen (écouter) at the door, but I didn't understand (comprendre) what I heard (entendre)" (Paraphrase of Chion, 1983). While most listening involves all four modes at once, it is at the *comprendre* level that the emotional qualities are assigned to sounds, derived from sounds, or represented by sounds. Michel Chion combines these listening modes into three:
- Causal Listening: "consists of listening to a sound in order to gather information about its cause (or source)"(Chion, 1990, p. 25).
- Semantic Listening: this involves listening to the "codes" within a sound and finding the meaning therein.
- Reduced Listening: "takes the sound-verbal, played on an instrument, noise or whatever—as itself the object to be observed instead of as a vehicle for something else" (Chion 1990, p. 29). This is a phenomenological approach to the experience of sound that reduces a sound to its essence and investigates its qualities (timbre and harmonic content, form, volume, etc.).

Chion's "semantic listening" relates most to Schaeffer's "comprendre," and it is this listening mode with which we decipher the emotional meaning encoded within a sound. Expanding on David Huron's six-component

theory of auditory-evoked emotion, Kai Tuuri, Manne-Sakari Mustonen, and Antti Pirhonen devised a system to understand the combination of the psychological and the philosophical modes of listening:
Preconscious modes:
- Reflexive: reflexive responses triggered by sound.
- Connotative: freely formed associations immediately evoked in listening.

Source-orientated modes:
- Causal: listening for the cause of a sound.
- Empathetic: listening for emotion or state of mind expressed through sound.
- Critical: critiquing a sounds suitability for a particular situation.

Quality-oriented mode:
- Reduced: objectively describing the properties of a sound.

In order for sound in digital media to make an emotional impact, an awareness of both the psychological and physiological immediacy of auditory awareness, and the philosophical, semiotic modes of listening and understanding the possible meanings conveyed with sounds, is required. From a design standpoint, engineers, programmers, and other creators of digital media keep both viewpoints in mind when creating sounds with emotional functions and their corresponding visual and haptic counterparts. Designing effective auditory feedback requires an understanding of auditory processing and the perceptual limitations of sound perception for users; for example, a designer might ask: "Is the sound loud enough for the user to hear? Is the pitch too low or too high to easily hear? Are important feedback sounds somehow hidden by louder, more trivial mechanical sounds produced by the device in question? Are sounds emitted too close to one another, causing them to be temporally masked?" Likewise, designers must also ask deeper philosophical questions: "What do the sounds of these instruments represent? Do they represent the same thing in every culture? At what point does a sound become noise, and when does a sound pass the point of being an alert to being a nuisance? Can a user easily derive the meaning from this sonic feedback alone, or are other sensory cues (visual, haptic, etc.) required?"

In his book, *Emotional Design: Why We Love (or Hate) Everyday Things*, Donald A. Norman discusses his research on the emotional impact of esthetics and design, concluding that emotions and cognition are intertwined. According to Norman, when a user encounters a designed product, we process our experience through three levels of perception: *visceral, behavioral, and reflective*. A user first experiences design on a visceral level, which includes affective reactions and emotions, and this is the most primal and reactionary form of perception wherein we decide if the design is good or bad, pleasing or disgusting, safe and comforting, or dangerous, or alienating, etc. This level of experience could correspond to Schaeffer's "*ouïr*" and/or "*écouter*" modes of listening, to Chion's "causal listening," and to Tuuri, Mustonen, and Pirhonen's preconscious modes of listening. Experience of design on the behavioral level is triggered by our reactions on the visceral level, and like the visceral level, is mostly unconscious; it describes how users act and feel in using the design, and whether or not using the design creates a meaningful experience, corresponding to Schaeffer's *entendre* and Tuuri, Mustonen, and Pirhonen's "causal" mode of listening. Last, reflective processing refers to the cognitive processing of a design and the rationalization of choices made as a result of the design. In this level of processing, users seek to understand a product, assess its value or ascribe value to it, even to the point of integrating the design into the expression of self-image, to have pride in the ownership of the product, and to attribute some cultural value to the design. This reflective level parallels Schaeffer's "*comprendre*," Chion's "semantic listening," and Tuuri, Mustonen, and Pirhonen's "empathetic" and "critical" modes of listening.

Our experience with a designed product begins with visceral and affective processing. Our first experience with design is marked by our emotional reaction to it, and our continued interaction depends upon whether or not the design affected an appropriate, desirable emotional response. Sound can reinforce or negate the emotional content designed into the visual elements of digital media, affecting our behavior. For example, a love song that plays in the background as star-crossed lovers meet for the first time makes a movie scene all the more saccharine, resulting in more emotional engagement, tears, etc.; likewise, horror movies often employ anempathetic sounds or music that incongruously signals stasis, calmness, or even happiness while a brutal murder is occurring within the visual *mis-en-scène* and is exploited to produce an additional feeling of uneasiness. Similarly, smartphone games with repetitive music or applications that produce sonic alerts that are too loud or those that occur too often are frequently countered with the phone's mute button. Upon hearing a particular ringtone, listeners can exercise their reflective perception, making value judgments about the brand of phone that plays such a ringtone, about the person who would own that brand of phone, and even compare this evaluation to past experiences with the same ringtone—it can remind a listener of happier times because this particular ringtone plays each time a loved one calls, or it can be an annoyance to an innocent bystander who is distracted from his or her work a little more each time the ringtone resounds loudly across the office.

(AUDITORY) DISPLAYS OF EMOTION

Rather than simply relying on one sense at a time, we interact with the world using many senses simultaneously. The combination of visual and aural feedback in digital media gives us a great deal of information, especially at the site of the human-computer interface. In an essay on nonspeech auditory output, Stephen Brewster lists several reasons why sound is beneficial in HCI (Brewster, 2007, p. 249). Vision and hearing are interdependent, and our ears signal to our eyes that there is something that demands visual attention; the temporal resolution of our auditory system is superior to that of our visual perception. Also, sound reduces the overload from large displays, so rather than bombarding users with tons of visual information that can easily be overlooked, some of that information is presented as sound instead. Sound also reduces the amount of information needed on screen and reduces the demands on a user's visual attention. While attending to a task that requires visual attention, one can rely on aural feedback to monitor the progress of other; for example, downloads and other more time-intensive processes are often assigned a sound that users hear whenever the process is complete. Meanwhile the user can attend to other business and wait for the signal, rather than constantly checking to visually confirm whether or not the download is complete. Similarly, when buying groceries at the supermarket, the cashier does not need to continually check the screen on the cash register to make sure an item was properly scanned; rather, the "beep" emitted from the machine provides aural confirmation that the item is accounted for, and the cashier can quickly continue to sell groceries. Brewster also notes that the auditory sense is underutilized, that sound is attention grabbing, that some objects or actions within an interface may have a more natural representation in sound, and that making computers more usable by the visually impaired is among the benefits of using sound in HCI design contexts[4] (Brewster, 2007, p. 249). While Brewster does not mention the emotional connections and connotations associated with sound and music, these are certainly assets that sound brings to HCI and digital media.

---

[4] Quoting G. Kramer, Brewster also lists some problems with using sound in HCI: the relatively low resolution of sound compared to high resolution visuals, presenting absolute data with sound is difficult, there is a lack of orthogonality and changing one attribute of one sound could affect the others, sound is temporal and information represented with sound is transient, and finally, some people easily find sonic feedback annoying (Brewster, 2007, p. 249).

Auditory display is the general term used to describe the ways in which sonic feedback from an interface communicates information to the user in HCI. The process of representing or perceptualizing data or other information using sound is called *sonification*. Reading printed information orally is an easily recognizable example of sonification—the information on the page is transduced from print (perceived visually) into intelligible speech (perceived aurally). Auditory displays are used within an interface to perform a number of functions, all of which communicate information to the user. According to Bruce Walker and Michael Nees (and many others including Buxton, 1989; Edworthy, 1998; Kramer, 1994), auditory displays perform alerting functions (esp. notifications and warnings), status and progress indication functions, data exploration functions (which is most applicable in the sonification of scientific data), and in art and entertainment applications (for the creation of computer music, digital sound art, immersive exhibitions, games, etc.) (Walker & Nees, 2012, p. 4).

Auditory displays employ several types of auditory feedback that are effective intermediaries of emotional information. *Auditory icons*, an idea developed by Gaver (1989, 1997), are the aural equivalent of visual icons and rely on their identifiablity from the analog world to transfer meaning to the digital domain. Visual icons are part of the graphical user interface that helps users to understand and interact with an electronic device; for example, users recognize the image of a trashcan on their computer desktop and understand that dragging a file to this icon presumably results in "trashing" the file, essentially deleting it, and freeing space on the computer's hard drive. The sound of crumpling paper the user hears as he or she drags a file into the trashcan icon is an auditory icon—it represents the act of discarding a file similar to the visual depiction of the trashcan.[5] Any emotional connection to a real-world sound can presumably transfer into HCI contexts if the sound is used as an auditory icon. Some auditory icons, such as the "clicks" that suggest a user is typing a text message on a smartphone, or the shutter sound that is used to indicate a picture is taken with a smartphone camera, are skeuomorphic[6] and are not necessary, per se (as any other sound could

---

[5] On computers that run Windows operating systems, the trashcan is exchanged for a more ecologically friendly "recycling bin." When it is utilized, users can hear the sound of a crushed aluminum can being thrown into an empty bin.

[6] A skeuomorph is a design element, usually ornamental, that mimics design qualities of older versions; digital skeuomorphs mimic design features of physical/analog objects. The tiny ring on the neck of some syrup bottles, the metal rivets on jeans, and the turning pages of e-books are all skeuomorphs.

replace them, or they could still function just as well with no sound at all), but they give users aural feedback that the interface is responding to his or her input, resulting in ease of use, positive action reinforcement, and ultimately results in positive emotional interactions with the interface for the user. In fact, skeuomorphic auditory icons are often used in digital media with the hope to elicit positive feelings of nostalgia in users. The camera shutter click on their camera phone reminds the user of the good times he or she had with their analog camera, the sound of a record ending that is artificially added to the end of some CDs hearkens back to the "good old days" of long play records (LPs), and users can choose ringtones that sound like an old-fashioned telephone for old times' sake. Apple iPhones and other smartphones also employ auditory icons that signify safety and security; when the phones are locked, users are reassured by the sound of a clicking lock that their data is secure. Digital security systems also reassure users that the system is armed and that their house is satisfactorily protected through voiced announcements or simple beeps.

"*Earcons*" are similar to auditory icons, but rather than being concrete, real-world sounds transposed into a digital interface, earcons are abstract, sometimes musical in nature, and are manufactured to present or represent information aurally to the user. Sonic branding and sonic logos (sometimes called "sogos") are perhaps the most famous types of earcons; in the USA, the three-tone NBC sonic logo, Intel's 5-tone sonic logo, the chord an Apple computer plays as it starts up, all instantly represent their corresponding brand to the listener. Unlike auditory icons, however, earcons are not everyday sounds that are mapped onto functions of the interface, so rather than being intuitively understood, the meaning of earcons must be learned with the system. Emotional responses to earcons are forged by emotional connections to brands, and hearing these earcons can remind listeners of this connection.

Earcons can easily be found on personal computers; each brand of computer has its own idiosyncratic (and trademarked) set of earcons that are mapped onto a wide array of tasks. The Windows operating system has a number of earcons specific only to Windows; some are replaced with each new version of Windows, and others remain for several software generations. Each version has an identifiable, musical start-up and shutdown earcon, reassuring users that the computer has been turned on and off safely.

Windows also uses a "tada" fanfare that signals completed tasks, a sound similar to a bubble popping that is played to indicate the "popping up" of a notice of various sorts, and "dings" and error alerts (an interval of a perfect 5th played in the upper registers of a piano) notify the user something is wrong or demands their attention. These sounds help to trace out acceptable parameters for computer usage, and they afford users a sense of familiarity and offer encouraging reinforcement when heard.

Simple *speech* is also used as auditory feedback in some HCI. America online (AOL) users were greeted with "You've got mail!" whenever they received an e-mail. Many phone systems and other voice user interfaces (VUIs) afford various levels of interaction with users in the form of spoken, aural information or a series of prompts, and he or she responds vocally or with the telephone keypad. While some companies rely on recordings of real-life speakers, others rely on synthetic samples of speech. In both cases, characteristically pleasant and calming voices are chosen for their esthetic and affective qualities. Similarly, "*spearcons*," a *portmanteau* of "speech" and "icon," were created by Bruce Walker, Amanda Nance, and Jeffrey Lindsay by speeding up a spoken phrase until it is not recognizable as speech; in testing their usefulness, the creators claim that "spearcons and speech-only both led to faster and more accurate menu navigation than auditory icons and hierarchical earcons … These results suggest that spearcons are more effective than previous auditory cues in menu-based interfaces, and may lead to better performance and accuracy, as well as more flexible menu structures" (Walker, Nance, & Lindsay, 2006, p. 63). Of course, speech can be used to convey emotion, and engineers are perpetually working to refine synthetic voices into more responsive, "emotive" utterances.

Furthermore, ambient sound and environmental noise can lead to immersive and affective experiences in digital media, particularly in video games, virtual reality, training simulations, film, and various forms of entertainment media. In digital media, the affective almost always trumps the objective. In film, nothing is "real," but rather, "hyper-real"—even time and space. Scriptwriter and film producer Jon Boorstin writes:

> *Philosophers may theorize about subjectivity, but working filmmakers try to learn exactly what it means to say that time is flexible, a function of our inner clock. They study how long a second really is, and how short, and what makes it feel one way or the other. They know that what we see isn't really what's out there because they've learned how spatial perception varies with angle and focal length and lighting, how "true" colors are a figment of lighting, and context, and even the glass of a lens. They know there is no "real" sound but only a better or worse approximation of what our ear expects.*
>
> **(Boorstin, 1990, p. 198)**

Because our experience of this digital media is subjective, the onus lies on the director to affect each subjective experience to elicit the desired emotional response. For example, consider the sound design in film, especially the documentary genre. Since a documentary film is not a piece of journalism, there is no ethical mandate to report, record, or reproduce the sounds that were actually heard at the time the footage was shot. Filmmakers add extra sound effects and ambience, take out sounds, attenuates noises, fills silences with meaningful sounds, and insert nondiagetic mood music at critical points in their story, all in the service of creating a "hyper-real" soundscape that moves the audience emotionally. In many video games, immersion is critical, and players must feel as if they are part of the game in order to succeed. Ambient sound is added to represent the virtual world that is navigated by the game player's avatar, and the sound from the avatar's footsteps, gunfire, sword strikes, and other actions—even the character's breathing—are processed in such a way that they are made to be perceived as if those actions are taking place within that specific environment. For instance, if the avatar is in a cave, almost all sound effects are processed with copious amounts of reverb; if the avatar is outdoors on a battlefield, no reverb is added, accurately matching the acoustic conditions of real life and producing an aurally immersive experience with the hope that players will also become emotionally immersed within the game. It is important to remember, too, that the creation of these types of hyper-reality requires a high degree of fluency with other forms of digital media technology and software that are likewise pervaded with aural and visual feedback designed activate emotional responses (especially amusement/engagement and trust) in the user.

Much like the way in which electric cars were required to add sound because they were too quiet (see footnote 1), extra sound is sometimes added to an interface to give users the feeling that the system is still responsive and functioning properly. For instance, Skype calls, cellular telephones, and digital radio stations use "comfort noise," that is, extra static or other synthetic background noise to fill silences in the transmission; without this, silences sometimes cause users to believe the call is dropped and they will hang up prematurely thinking the transmission has failed and that they have been somehow disconnected. "Music on hold" is another example of this supplementary sound. Callers who are placed on hold will frequently hear "Muzak," "elevator music," recordings of classical music or jazz, or sometimes even a live feed from a terrestrial radio broadcast as they wait "on hold"; this music is supposed to give the listener a sense of assurance and comfort that he or she has not been disconnected.

"Extra" sound is also sometimes used to make technology more comfortable. "Siri" and other "intelligent personal assistants," or the vocal feedback provided by global positioning systems (GPS) navigation interfaces are all designed to be comforting and familiar voices with which users can interact. Advanced HCIs, namely artificial intelligence and advanced robotics, are sometimes eerie and disconcerting to humans, especially when the interface is humanoid. Computer voices have long been considered disturbing; HAL 9000 from Stanley Kubrick's film *2001: A Space Odyssey* (1968) spoke with an obviously unnatural, monotone, dispassionate voice—totally computer yet somehow still human. In the 1970s, Masahiro Mori developed the concept of the "Uncanny Valley," an esthetic theory that explains the human tendency to react with revulsion toward robots that were similar, but not quite identical, to humans (Mori, 1970/2012). As technology advances and robots continue to look, move, and act more like humans, this particular issue becomes ever more poignant, especially when these robots are being used for medical purposes and other critical social situations. To address this issue, sound libraries are being developed and implemented to help robots express intention and emotion in social interactions, thereby making them more familiar and less unnerving to humans. Others are conducting research to discover how far voice anthropomorphism can be stretched without causing unwanted emotional reactions in humans, and to discern what types of voices are best suited for emotional robot-human interactions (Cowan, 2014; Niculescu, Dijk van, Nijholt, & See, 2011; Riek, Rabinowitch, Chakrabarti, & Robinson, 2009). Sound is an important part of affective computing and has the potential to play a large part in bridging the uncanny valley.

CONCLUSION: SOUNDING SAFETY, SECURITY, STASIS, AND STATUS

Consumers tend only to purchase and continue to use products that cultivate positive feelings. Companies spend countless hours and untold amounts of money on focus groups to ensure potential customers feel happy using their product. Users rely on both positive and negative feedback from digital media to confirm whether or not they are using the interface properly. Sound plays a critical role in assuring users that they are utilizing digital media effectively and correctly, comforting them in the knowledge that they are safely using a product. Sound also signifies status for users of digital media and helps to formulate to various degrees an emotional sense of self-worth; whether a user is proudly staking claim over his or her environment by playing music loudly from a smartphone, bragging about a recent purchase of the newest phone by conspicuously leaving the ringtone on for all to hear, or by sharing nostalgic feelings and tastes in music through playing a music video game, the sonic elements of digital media make an emotional impact on the user. As technology marches toward more ubiquitous computing and the "Internet of Things," wherein users might simply use his or her voice to control household appliances, a car, and almost all other electronic devices, and as anticipatory computing develops, allowing your computer to fetch information you need before you even request it, it is imperative that the emotional impact of the sounds associated with this technology, and the ways in which emotion is communicated through these sounds, is better understood.

REFERENCES

Abeles, H. F., & Chung, J. W. (1996). Responses to music. In D. A. Hodges (Ed.), *Handbook of music psychology* (2nd ed., pp. 285–342). San Antonio, TX: IMR Press.

American Speech-Language Hearing Association. (1996). Central auditory processing: Current status of research and implications for clinical practice. *American Journal of Audiology*, 5(2), 41–54.

Attali, J. (1985). *Noise: The political economy of music*. Minneapolis, MN: University of Minnesota Press.

Baker, D. (2013). Did you know BMW's door click had a composer? It's Emar Vegt, an aural designer. *Wired Magazine*, Retrieved from http://www.wired.co.uk/magazine/archive/2013/04/start/music-to-drive-to.

Boorstin, J. (1990). *Making movies work: Thinking like a filmmaker*. Los Angeles: Salman-James Press.

Brewster, S. (2007). Nonspeech auditory output. In A. Sears & J. Jacko (Eds.), *The human-computer interaction handbook: Fundamentals, evolving technologies and emerging applications*. (2nd ed.). Boca Raton, FL: CRC Press.

Buxton, W. (1989). Introduction to this special issue on nonspeech audio. *Human-Computer Interaction*, 4, 1–9.

Chion, M. (1983). *Guide des object sonores, Pierre Schaeffere et la researche musicale* (J. Dack and C. North, Trans., 1995). Paris: Ina-GRM/Buchet-Chastel.

Chion, M. (1990). *L'audio-vision. Paris: Nathan [Audio-vision. Sound on Screen]*. New York: Columbia University Press. C. Gorbman, trans.

Cowan, B. R. (2014). Understanding speech and language interactions in HCI: The importance of theory-based human-human dialog research. In *Designing speech and language interactions workshop, ACM conference on human factors in computing systems, CHI 2014*. Available online, http://www.cs.bham.ac.uk/~cowanbr/understanding%20speech%20and%20language%20interactions%20in%20HCI.pdf.

Edworthy, J. (1998). Does sound help us to work better with machines? A commentary on Rautenberg's paper 'about the importance of auditory alarms during the operation of a plant simulator. *Interacting with Computers*, 10, 401–409.

Feld, S. (1996). Waterfalls of song: An acoustemology of place resounding in Bosavi, Papua New Guinea. In Steven Feld & Keith Basso (Eds.), *Sense of place*. Santa Fe, NM: School of American Research Press.

Gabrielsson, A. (2001). Emotions in strong experiences with music. In P. N. Juslin & J. A. Sloboda (Eds.), *Music and emotion: Theory and research* (pp. 431–449). Oxford: Oxford University Press.

Gaver, W. (1989). The sonicfinder: An interface that uses auditory icons. *Human Computer Interaction*, *4*(1), 67–94.
Gaver, W. (1997). Auditory interfaces. In M. Helander, T. Landauer, & P. Prabhu (Eds.), *Handbook of human-computer interaction* (2nd ed., pp. 1003–1042). Elsevier: Amsterdam.
Jackson, D. M. (2003). *Sonic branding: An introduction*. New York: Palgrave Macmillan.
Juslin, P. N., & Västfjäll, D. (2008). Emotional responses to music: The need to consider underlying mechanisms. *Behavioral and Brain Sciences*, *31*(5), 559–575.
Kisilevsky, B. S., Hains, S. M. J., Lee, K., Xie, X., Huang, H., Ye, H.-H., et al. (2003). Effects of experiences on fetal voice recognition. *Psychological Science*, *14*(3), 220–224.
Kramer, G. (1994). An introduction to auditory display. In G. Kramer (Ed.), *Auditory display: Sonification, audification, and auditory interfaces* (pp. 1–78). Reading, MA: Addison Wesley.
LaBelle, B. (2008). *Background noise: Perspectives on sound art*. New York: Continuum.
Longhi, E., Pickett, N., & Hargreaves, D. J. (2015). Wellbeing and hospitalized children: Can music help? *Psychology of music*, *43*, 188–196.
Lyman, B., & Waters, J. C. (1989). Patterns of imagery in various emotions. *Journal of Mental Imagery*, *13*, 63–74.
Meyer, L. B. (1956). *Emotion and meaning in music*. Chicago, IL: University of Chicago Press.
Mori, M. (1970/2012). The uncanny valley (K.F. MacDorman and N. Kaggeki, Trans.). *IEEE Robotics and Automation Magazine*, *19*(2), 98–100.
Niculescu, A., Dijk van, B., Nijholt, A., & See, S. L. (2011). The influence of voice pitch on the evaluation of a social robot receptionist. In *Proceedings of the second IEEE international conference on user science and engineering, I-USEr2011, 29 November–2 December 2011, Kuala Lumpur, Malaysia* (pp. 18–23).
Osborne, J. W. (1980). The mapping of thoughts, emotions, sensations, and images as responses to music. *Journal of Mental Imagery*, *5*, 133–136.
Plutchik, R. (1984). Emotions and imagery. *Journal of Mental Imagery*, *8*, 105–111.
Riek, L. D., Rabinowitch, T., Chakrabarti, B., & Robinson, P. (2009). Empathizing with robots: Fellow feeling along the anthropomorphic spectrum. In *Proceedings of the IEEE international conference on affective computing and intelligent interaction (ACII '09), Amsterdam, Netherlands, September 10–12, 2009*.
Schaeffer, P. (1977). *Traité des objets musicaux*. Paris: Le Seuil.
Scherer, K. R. (1999). Appraisal theories. In T. Dalgleish & M. Power (Eds.), *Handbook of cognition and emotion* (pp. 637–663). Chichester: Wiley.
Sloboda, J. A. (1991). Musical structure and emotional response: Some empirical findings. *Psychology of Music*, *19*, 110–120.
Walker, B. N., Nance, A., & Lindsay, J. (2006). Spearcons: Speech-based earcons improve navigation performance in auditory menus. In: *Proceedings of the 12th international conference on auditory display, London, UK, June 20–23* (pp. 63–68).
Walker, B. N., & Nees, M. A. (2012). Theory of sonification. In T. Hermann, A. Hunt, & J. Neuhoff (Eds.), *The sonification handbook* (pp. 9–39). Berlin: Logos.

CHAPTER 3

Emoticons in Business Communication: Is the :) Worth it?

Jennifer M. Loglia, Clint A. Bowers
University of Central Florida, Orlando, FL, USA

A Vignette
Imagine starting a brand new day at work. En route to your desk, you greet co-workers as you pass by. You see Mike peering over his monitor in your direction, and you go to stop by and ask what's up. After answering the question he had, you arrive at your chair; you're about to say hi to your neighbor, Cathy, when you notice she's staring perplexingly at her notes. You decide that you can wait to say hey. In the meantime, you boot up the ol' computer and do what every office worker does first thing in the morning: check email. Sitting in your inbox is an email from your boss. You click it open, and it reads:
Meet me in my office at 11 am.
Oh no. What does that mean? Is this good? Bad? Are you getting reprimanded? Or worse, fired? Or, are you finally getting that raise that you keep hinting at and so clearly deserve? Your boss didn't give you much with which to work. If only you had some other kind of hint as to what she meant. The words are there, but they might as well be in Wingdings without context. But wait, how did you know that Mike had a question? Or that Cathy didn't want to be bothered? Neither of them said one word to you, but you figured it out anyway.
Let's say your boss emailed you this message instead:
Meet me in my office at 11 am :).
Assuming your boss isn't a sadist, you can safely speculate the interaction you will have will be a positive one. Those two extra symbols combined made all the difference. Using an emoticon, a pictorial representation of a facial expression, provides a wealth of information that text alone cannot supply.

INTRODUCTION

Online communication is extremely prevalent in our current society, and it is an essential business tool. Almost everyone participates in some form of computer-mediated communication (CMC). It allows us to communicate with a person halfway around the world or a person in the cubicle next to us, send a message that can be answered later, or one that needs an immediate response. And, although CMC is very convenient, it's arguable when it comes to efficiency. CMC is efficient in that it is quick and not constricted by the bounds of time or location. But, we lose layers of conversational depth found in face-to-face (FTF) communication when we communicate electronically, and that could lead to unintended interpretations and outcomes. Even video conferencing is not a complete substitute for FTF communication (Nguyen, 2008). Therefore, CMC may not always be as "efficient" as it seems.

Nonverbal cues are incredibly important (Frith, 2009; Riggio, 2005). They offer valuable information such as tone and context. We hear voice inflection and volume level. We see facial expressions, hand gestures, and body language. They help us differentiate that genuine and happy "thanks!" from that scathing and sarcastic *"thanks."* We attempt to insert our own form of these cues in CMC within the text itself (as I just attempted with my exclamation point and italics) and with the use of emoticons. While these are helpful in conveying the missing communicative data points, we are taught early on in our professional grooming that emoticons and the like are unprofessional (Seaton, 2011). We are heavily discouraged from using these useful tactics in any formal email communication. But why?

Emoticons especially are regarded as "childish" (Emerson, 2011). One reason they are regarded as such could be because they can appear cartoonish. Another reason could be because they are more popular and more often used among younger populations ("millennials"). Maybe those who hold the highest standards of "professionalism" feel threatened, viewing emoticons as a gateway for other "unacceptable" forms of corrupt communication, such as "text-speak" (e.g., "u r" or "brb"). Regardless of the reasoning, the outcome is the same. What can we do to change that?

One approach is to take advantage of new technologies. Communication is an important aspect of the workplace, especially as our workplace becomes more virtual. Increasing emphasis is placed on the virtual office, and managers are faced with the challenging question "how can I manage them if I can't see them?" (Cascio, 1999, p. 1). With this change in structure comes the need to respond in an appropriate, effective manner. One workplace theory that has been linked to many positive outcomes is leader-member exchange (LMX) (Gerstner & Day, 1997). LMX focuses on the quality of the relationship between supervisor and subordinate. Uhl-bien (2003) noted that relationship development may be a key factor that makes for great leader-subordinate relationships, and communication is a fundamental relational skill. Mayfield and Mayfield (1996) showed that subordinates who had higher quality relationships with their leaders outperformed their peers with lower quality leader relationships by about 20% and were about 50% more satisfied. Therefore, we should be welcoming methods that improve and clarify communication, not turning them away. This could affect tangible outcomes and improve businesses as a whole. Since we understand the role of nonverbal cues in FTF communication, these same properties should translate to emoticons in CMC, and we should be able to achieve better communication, leading to increased LMX.

NONVERBAL COMMUNICATION

Nonverbal cues are those that we perceive outside of what is actually being said to us (either verbally or in text). Examples of nonverbal cues are differences in voice tone, facial expressions, and hand gestures. The main functions of nonverbal cues are to convey emotions and attitudes and to "emphasize, contradict, substitute, or regulate verbal communication" (Wei, 2012, p. 2, as cited in Jibril & Abdullah, 2013, p. 201). We try to seek out these nonverbal cues to provide us with more information. How does this happen? Facial expressions are a prime example of how these behaviors can provide communicative information. Frith (2009) provided an example:

> At first, the facial expression of fear has direct behavioural advantages for the actor, since widening the eyes for example, increases the visual field, thereby increasing the likelihood of detecting signals of danger. This expression then becomes public information that observers can use as a signal to be vigilant. In the next step, the actor becomes able to control the sending of a signal that was previously emitted inadvertently. Through such control he can express sorrow and embarrassment as a means of appeasing aggression in others. Finally, both the actor and the receiver become aware that they are exchanging signals and that these can be used for deliberate communication. At this stage, the signals need no longer be tied to their original behavioural function. They can be arbitrarily related to meaning, making the development of language possible (p. 3457).

Research has acknowledged the lack of nonverbal cues in CMC and the disruption it causes in communication (Kiesler, Siegel, & McGuire, 1984; Walther, 1992; Walther, Slovacek, & Tidwell, 2001). Emoticons are filling that missing gap as CMC's nonverbal cues by simulating nonverbal emotional behaviors. Social Information Processing Theory (Walther, 1992, 1996) states people will try and compensate for missing communication cues via proxy cues. Channel Expansion Theory (Lo, 2008) explains how people learn how to use emoticons when communicating through CMC; Channel Expansion Theory states that, after communicating in a certain medium for some time, the user learns specific skills and gains knowledge to assist communication in that medium. So, it is not surprising that emoticons have made their way into all kinds of CMC, and that includes CMC that's used at the office.

Nonverbal communication has been studied in the context of the workplace and in leadership. Three things are notable in this literature: one, feedback in a F2F situation is more positive and accurate (Hebert & Vorauer, 2003), two, even though CMC lacks generally lacks verbal cues, emoticon still comes through this medium (Byron, 2008), and three, nonverbal communication is an important part of leadership (e.g., Burgoon, Buller, & Woodall, 1995, as cited in Teven, 2007). We will now explore each of these findings in detail to further understand why nonverbal communication is important in the workplace.

Several studies have found that feedback in a F2F situation is more positive and accurate and can influence outcomes in different ways. Nonverbal cues significantly affect the results of interview processes (e.g., Chapman & Rowe, 2001; Connerley & Rynes, 1997; Liden, Martin, & Parsons, 1993). For example, Liden et al. (1993) found that interviewers' body language affected how well applicants did as they were rated by objective judges; applicants who had interviewers with "warm" body language did better than those applicants whose interviewers appeared "cold." Other feedback is also affected by nonverbal cues and CMC. Hebert and Vorauer (2003) found that F2F evaluations were more positive than CM evaluation and, when referring to task-relevant information, more accurate. They concluded that these results were due to the receivers' access to nonverbal cues; these cues provided a higher quality of evaluation. Kurtzberg, Naquin, and Belkin (2005) found that employees gave more negative evaluations of their peers over email compared to those given in paper form. This was mediated by the fact that employees felt a lesser sense of social obligation when filling out email evaluations. This study did not look at FTF evaluations. Whether the ratings were more accurate was unknown. Potentially, the email evaluations could have been more accurate, offering a counter to Herbert and Vorauer, and making a case as to why email evaluations might be better for honesty. But, Herbert and Vorauer's case shows us that F2F (and therefore nonverbal cues) help give a more positive feel to evaluations while still remaining accurate. We know that employees feel that it is difficult to portray positivity in emails (Markus, 1994), especially due to the nature of what they are talking about (e.g., tasks) and in the manner in which they confer (e.g., with a "serious" tone) (Lea & Spears, 1992). Even happier messages are often dulled by the lack of nonverbal cues (Byron, 2008). Emoticons can be that missing link, taking advantage of the best parts of both communication mediums, and playing the role of nonverbal cues in CMC. The evaluation process is an important part of the supervisor-subordinate relationship and represents an area where a supervisor can energize and motivate their employees, so having a positive impact is important (Riggio, 2005).

Despite the lack of cues allowing for emotional information in CMC, emotion is still passed through CMC, leading to miscommunication, and there's little research on how or why this happens (Byron, 2008). In a study by Byron and Baldridge (2005), a focus group reported expressing and receiving emotions in work email, and almost all of them said they had issues doing so. This same focus group also listed contradictory or disagreeing ways to express positive emotion in emails (e.g., sending longer or shorter messages, using or not using exclamation points). Clearly there is no consistency in "emotional typing" and it is causing miscommunication. For example, person A could write a one-sentence email to person B, and person B could interpret the short email as person A being mad or upset, when in reality, person A only had a small amount of time to write the email and is not mad at all. Emoticons, which mimic facial expressions, are a much more consistent (albeit not perfect) system of nonverbal communication. Their use could (and arguably do) more easily portray positivity (and other emotions) and decrease the misinterpretation of plain text.

Nonverbal communication also has an impact on leadership, and we need to better understand the nonverbal communication possible in CMC. Burgoon et al. (1995) (as cited in Teven, 2007) acknowledged nonverbal behaviors as important to the leader-member relationship, because they are important and central to interpersonal interactions. Ilies, Curşeu, Dimotakis, and Spitzmuller (2013) found that leadership outcomes are related to emotional expressiveness via idealized influence; they also found that relational authenticity was very important when it came to influencing others. Exchanges of emotional expressiveness between leaders and members led to higher idealized influence, which affected members' perceptions of leader effectiveness and the reported effort members put forth in their work. Recognizing the emotional part of leadership does not only address the "affect" part of LMX (which you will find outlined in more detail later), but also addresses transformational leadership. Idealized influence (a.k.a. charisma) is one of the four areas that comprise transformational leadership (Purvanova & Bono, 2009). Transformational leadership has been documented as an important part of virtual teams (Avolio, Kahai, & Dodge, 2001; Ruggieri, 2009). It has been shown that those leaders that engage in transformational leadership achieve higher team performance levels (Purvanova & Bono, 2009). Teven (2007) also found nonverbal immediacy, described as a mix of behaviors that enhance the perceived closeness in relationships (Mehrabian, 1969), helps to increase affect between supervisors and subordinates, and leads subordinates to have more credible perceptions of their supervisors. Other research has shown that when leaders hold positive expectations of their employees, worker productivity increases; these expectations are demonstrated both verbally and nonverbally (Eden, 1984, 1993; Eden & Shani, 1982). One common theme in all of this research is the role that nonverbal cues play in increasing the effectiveness of leadership. Again, nonverbal cues are providing more information and context to message receivers (in this case, subordinates), and in turn, leaders can appear more authentic, relatable, charismatic, and closer to their subordinates. As we can see through the literature, the opportunities to express nonverbal cues in CMC in the business setting are very limited, and we are potentially limiting leader effectiveness. If accepted in this setting, emoticons can be these missing cues and allow beneficial information to flow from one party to another.

EMOTICONS

Emoticons have been defined by many researchers in the past decade. Rezabek and Cochenour (1998) defined emoticons as "visual cues formed from ordinary typographical symbols that when read sideways represent feelings or emotions" (p. 201). Walther and D'Addario (2001) described them as "graphic representations of facial expressions" (p. 324). Danesi (2009) said emoticons are a "string of keyboard characters that, when viewed sideways (or in some other orientation), can be seen to suggest a face expressing a particular emotion" (p. 110). Examples of emoticons can be found in Table 3.1.

The literature is starting to differentiate between emoticons and "smilies." Ganster, Eimler, and Kramer (2012) identify emoticons by "character strings" and smilies as "graphical pictograms." A smilie looks like this ☺ or ☹ (it's interesting to note that Microsoft Word 2010 and 2013 automatically

Table 3.1 Western emoticons

Emoticon	Meaning
:), :-), =), :], =]	Smiley, happy face
:(, :-(, =(, :[, =[, :<, :C	Frown, sad face
:D, =D, :-D, XD	Big grin, laughing
:P, =P, >P	Tongue sticking out, cheeky/playful, blowing a raspberry
>:(, >(	Angry
:/, =/, :\, =\, :-/, =-\	Indifferent, uneasy, skeptical, hesitant
>:0, :O, =o, =-O	Surprise, shock
;), ;-), ;D, ;]	Winking face, smirk

Note: Adapted from "List of emoticons."

turn an emoticon smile ":)" into a smilie "☺"). In that same study, Ganster et al. found that there is no difference in message interpretation between emoticons and smilies, but smiling smilies end up having a stronger impact on personal mood and therefore "elicit a stronger impact than emoticons" (p. 226). This is an important difference to note since there are more and more chat and messaging programs where these types of emoticons are standard.

It is also good to briefly acknowledge the cultural differences of emoticons and emoticon usage. The "sideways emoticons" are more popular in western cultures, but eastern cultures, such as Japan, have a completely different set of emoticons (in Japan, known as *kaomoji* (Katsuno & Yano, 2007)) and have different cultural references and usage (Katsuno & Yano, 2002). Examples of eastern emoticons can be found in Table 3.2. Many eastern

Table 3.2 Eastern emoticons

Emoticon	Meaning
^_^, (^__^), ^o^	Happy, normal laugh
;-;, ;_;, T_T, Q.Q, (ToT)	Sad, crying
^^;, ^_^;, (^^;)	Nervous, embarrassed, shy, sweat drop/cold sweat
^.~, ^_~, ^_-	Wink
(°_°), o.O, O_o	Confused
>_<, (¬_¬), o(-'д'-。)	Angry, annoyed
(╯°□°)╯ ︵ ┻━┻	Table flip—in this emoticon (read right to left), a person is flipping a table (used to express anger or frustration)

Note: Adapted from "List of emoticons."

emoticons are read from top to bottom vs. rotated 90 degrees. Facial *kaomoji* also have much more emphasis on the eyes, whereas western emoticons focus a lot on the mouth (Yuki, Maddux, & Masuda, 2005). This divergence could be because people from the east focus disproportionally more on the eyes when determining facial expressions vs. people from the west, who focus on the eyes and mouth equally (Jack, Blais, Scheepers, Schyns, & Caldara, 2009). Because they use non-Latin characters, eastern emoticons can be much more complex than western ones, as you can see with the "table flip" example in Table 3.2. This is just to show that, not only may there be cultural differences surrounding emoticons and their use, but also that there are completely different sets of emoticons out there.

Derks, Bos, and Von Grumbkow (2007) stated that "emoticons resemble facial nonverbal behavior" (p. 379). Luor, Wu, Lu, and Tao (2010) emphasized the contraction of "emotion" and "icon" and how an emoticon is "a creative and visually salient way" to add emotion and expression into a text-based message (p. 890). As stated previously, Jibril and Abdullah (2013) summarized emoticon literature and concluded that emoticons are the nonverbal cues that text-based CMC is/was lacking. In some cases, emoticons are literally their FTF, nonverbal cue counterpart. For example, a smiling face is represented by ":)" which looks exactly like a smiling face (as interpreted in the west). So it's not a far-fetched idea to put a smiley face emoticon after a written CM sentence if you would smile during/after saying the sentence verbally. It helps the sender to communicate the same emotion(s) to the receiver of the message. Emoticons (and other nonverbal cues as well) can also be thought of as an illocutionary force. Illocution focuses on the intent of the speaker and not necessarily what is literally being said. It acts as a clue for the receiver of the message on how to interpret said message (Dresner & Herring, 2010). For example, in that same paper, Dresner and Herring stated:

> In the following public e-mail post to the AoIR mailing list, the winking smiley is used to indicate that the utterance that immediately precedes it is not intended as a serious summons of the (deceased) media scholar Marshall McLuhan, but rather as a joke:
> Paging Mr. McLuhan....;)
> The winking emoticon here is best conceived of as a sign of the force of what has been (textually) said, rather than as an indication of emotion (p. 256).

Emoticons, as established previously, are primarily used to express emotion in CMC. The same regions of the brain are used when we read emoticons as when we see nonverbal cues (Yuasa, Saito, & Mukawa, 2011). Emoticons have been found to be able to strengthen a message, clarify or emphasize feelings, soften a negative tone, and/or express humor (Derks, Bos, & Von Grumbkow, 2008; Derks et al., 2007). Lo (2008) has found that using an emoticon (compared to not using one) affected the reception of emotion, attitude perception, and attention of the receiver. Messages that utilize emoticons were perceived to be "richer" in information, and they allow the sender to be more efficient and effective in their message (Huang, Yen, & Zhang, 2008). But, Luor et al. (2010) found that the use of emoticons in unnecessary circumstances did not add anything to the message, which echoes Walther and D'Addario's (2001) findings that the emoticons' impact decreased with overuse or unnecessary use.

Luor and other's study is notable here because they studied IMs in the workplace. They found that "positive emoticons significantly enhanced emotion when…discussing and coordinating tasks," and therefore they suggested that "positive emoticons should always be employed in work coordination tasks, especially when there is a tendency for unpleasant emotions to be felt…" (p. 894). This, along with Lo's study, is significant because these results show that there is a change in the *receiver*, either within his/herself or how he/she reacts. Another study by Utz (2000) found that, in online multiuser dungeon (MUD) games (a game type that is usually text-based and has multiple users in real time), the use of emoticons and other game-related cues was a significant predictor of relationship formation.

Gender is a variable that has been studied when it comes to the use of, and impression made by, emoticons. There is mixed research as to which gender uses emoticons more often. Wolf (2000) and Lee (2003) both concluded that men used little to no emoticons when conversing with other men, but when conversing with women, men used more emoticons to match that of their counterparts. Although these studies and others (Luor et al., 2010) suggest that women use emoticons more, Huffaker and Calvert (2005) analyzed online blogs and found that men used more emoticons than women did. Wolf (2000) also found that women used emoticons mainly to express humor, whereas men used them to express sarcasm. Gender and masculinity/femininity has also been looked at in the context of the workforce, specifically personnel selection. Using emoticons was found to portray stereotypical feminine qualities such as warmth. This, in turn, lead applicants using emoticons to be perceived as less competent and lower in stereotypical masculine attitudes and behaviors (e.g., leadership) when applying to male-gender-typed jobs. This also led to a lower starting pay rate for those applicants using emoticons (Thompson, Mullins, & Robinson, 2010).

The literature has shown that emoticons do their job in conveying emotion and being the missing nonverbal cues in CMC. We interpret emoticons as facial expressions and are aware of their cultural and gender contexts. We see how emoticons can be beneficial in the workplace, and, understanding how nonverbal cues can increase effective leadership, we can understand how emoticons can increase effective leadership.

LEADER-MEMBER EXCHANGE

LMX is a unique measurement of leadership that focuses on the quality of the relationship between the leader and the member and recognizes that the relationships established between leaders and each individual member may not be identical (Dansereau, Graen, & Raga, 1975; Gerstner & Day, 1997; Graen, 1976; Graen & Cashman, 1975; Graen, Novak, & Sommerkamp, 1982; Graen & Scandura, 1987; Graen & Uhl-Bien, 1991; Graen & Wakabayashi, 1994; Liden & Graen, 1980). Scandura, Graen, and Novak (1986) defined LMX as "(a) a system of components and their relationships (b) in both members of a dyad (c) involving interdependent patterns of behavior and (d) sharing mutual outcome instrumentalities and (e) producing conceptions of environments, cause maps, and value" (p. 580). LMX is based off of role theory (Graen, 1976), and role theory stresses that roles are multidimensional (Jacobs, 1971; Katz & Kahn, 1978). Therefore, Dienesch and Liden (1986) proposed that LMX consisted of three "currencies of exchange." Later, Liden and Maslyn (1998) added a fourth dimension to LMX. Table 3.3 defines the four dimensions of LMX.

Liden, Sparrowe, and Wayne (1997) grouped the LMX antecedent variables into four categories: subordinate characteristics, leader characteristics, interactional variables, and contextual variables. Martin, Epitropaki, Thomas, and Topakas (2010) grouped LMX outcome variables based on Gerstner and Day's (1997) review into three categories: attitudes and perceptions, behaviors, and task performance.

Communication is important when building relationships. Nonverbal behaviors such as facial expressions and body movements are crucial to the leader-member relationship (Burgoon et al., 1995, as cited in Teven, 2007). This nonverbal immediacy (again, described as a mix of behaviors that enhance the perceived closeness in relationships (Mehrabian, 1969)) is a way for leaders to increase interpersonal affect with their subordinates (Teven, 2007). This may be because nonverbal cues give more information about the message and make the message clearer for subordinates, who in

Table 3.3 Dimensions of leader-member exchange

Dimension	Definition
Affect	The mutual affection members of the dyad have for each other based primarily on interpersonal attraction, rather than work or professional values. Such affection may be manifested in the desire for and/or occurrence of a relationship, which has personally rewarding components and outcomes (e.g., a friendship)
Loyalty	The expression of public support for the goals and the personal character of the other member of the LMX dyad. Loyalty involves a faithfulness to the individual that is generally consistent from situation to situation
Contribution	Perception of the current level of work-oriented activity each member puts forth toward the mutual goals (explicit or implicit) of the dyad. Important in the evaluation of work-oriented activity is the extent to which a subordinate member of the dyad handles responsibility and completes tasks that extend beyond the job description and/or employment contract; likewise, the extent to which the supervisor provides resources and opportunities for such activity
Professional Respect	Perception of the degree to which each member of the dyad has built a reputation, within and/or outside the organization, of excelling at his or her line of work. This perception may be based on historical data concerning the person, such as: personal experience with the individual; comments made about the person from individuals within or outside the organization; and awards or other professional recognition achieved by the person. Thus it is possible, though not required, to have developed a perception of professional respect before working with or even meeting the person

Note: From Liden and Maslyn (1998), p. 50.

turn may be able to relate more with their supervisor since they have a (perceived) clear message. Research has shown that students are more likely to engage with their instructors outside the classroom when nonverbal immediacy is present (Fusani, 1994). If leaders (especially more virtual leaders) are using positive emoticons (thus increasing their nonverbal immediacy), we could potentially see an increase in leader-member affect, and therefore LMX. Affect focuses on the "relationship" aspect of leadership, and we know that using nonverbal cues is important in building interpersonal relationships (Burgoon et al., 1995, as cited in Teven, 2007). As established previously, emoticons are the nonverbal cues of CMC, so it would not be surprising to see the properties of nonverbal cues in FTF communication translate to emoticons in CMC. Unfortunately, using emoticons so openly in the business setting may be hard to do.

Emoticons are considered unprofessional, and people are advised not to use them in a business setting (Seaton, 2011). Because "professional respect" impressions can be made even before meeting someone, and because they are considered unprofessional, there is a risk that the use of emoticons might decrease professional respect, but not so much as to counteract the increase in affect. Although emoticons are still considered unprofessional, they are more common to and more in-use by younger generations (Krohn, 2004). Therefore, the use of emoticons in today's setting may not have the strong impact others might expect due to their acceptance by younger generations and the younger workforce, but it still might be present. Although wildly popular and acceptable in almost any other setting, it is still taught today that emoticons are not acceptable in professional material.

There is not really research or evidence that shows how positive affect affects contribution or loyalty. Research has shown that when there are similarly higher levels of agreeableness (based on the Big Five) between leaders and subordinates, there are higher levels of contribution and loyalty between the two (Ryan, 2009). There is also evidence to show that coworkers who give social support receive it in return (Bowling, Beehr, & Swader, 2005). These studies show that intangible "goods" can be and are being exchanged in the workplace, and that positive affect (if correlated with, or of similar nature to, agreeableness) could be linked to contribution and loyalty when measuring LMX. But there is too little evidence to show that positive affect affects contribution and/or loyalty to draw conclusions as to how emoticons may affect LMX via these attributes.

A meta-analysis conducted by Gerstner and Day (1997) found "significant relationships between LMX and job performance, satisfaction with supervisor, overall satisfaction, commitment, role conflict, role clarity, member competence, and turnover intentions" (p. 827). Increasing LMX leads to better outcomes in all of these areas, goals that (theoretically) all organizations strive for. Therefore, being able to increase LMX is something that all organizations would want to do. Despite its stigma, emoticons may be a simple and easy way to do this. By utilizing the benefits of nonverbal cues through emoticons, we could increase leader-member affect and increase LMX, yielding more effective leadership and an overall better workplace.

With updated methods comes updated etiquette. The literature is heavily lacking investigations of emoticons in business communication. There are so many unanswered questions, and many different studies could address these. Correlational studies could be conducted to look at emoticon use and LMX (or general workplace relationships). These could see if there is a relationship between workplaces that tend to use emoticons in their CMC and higher quality relationships among employees. Experimental studies also offer themselves as a useful method to assess the effects of emoticons on business CMC. Email and instant messaging communication can easily be manipulated for content and context. Although it may lack fidelity, an experimental study has the advantage of being able to isolate the use of emoticons.

The acceptance of emoticons is sneaking its way in, whether it is due to a younger generation entering the workforce, employees discovering emoticons' usefulness, both, or neither. Since CMC is such a common method of communication, even though an email may be "business related," workers are differentiating between personal and/or "lower level" work-related emails (such as to a colleague) and emails that are sent to superiors or clients. It is more generally accepted to use a few emoticons here and there in emails to colleagues, but general consensus would *still* never use them in emails to superiors or clients. It is almost hypocritical to talk about how much body language plays an important role in business and business relationships, but then completely wipe out that aspect when referring to CMC. We know that smiling has many positive effects (Guéguen & De Gail, 2003) and would encourage employees to smile with clients F2F. Yet, we are saying the same does not apply when using CMC.

CMC has become a norm and a standard. Organizations should always be on the lookout for simple, effective ways to make improvements. The use of emoticons could be a useful approach to optimizing communication and clarifying intention. We understand the role of nonverbal cues in communication, and that they provide further information and clarity to the communication process. We understand that the emoticons are these nonverbal cues in CMC. And, we understand that communication can affect organizational outcomes as we can see with LMX. We need to move forward and let go of old principals in exchange for new ones – ones that offer beneficial, tangible results.

REFERENCES

Avolio, B. J., Kahai, S. S., & Dodge, G. E. (2001). E-leadership: Implications for theory, research, and practice. *Leadership Quarterly, 11*, 615–668.
Bowling, N. A., Beehr, T. A., & Swader, W. M. (2005). Giving and receiving social support at work: The roles of personality and reciprocity. *Journal of Vocational Behavior, 67*(3), 476–489.
Burgoon, J. K., Buller, D. B., & Woodall, W. G. (1995). *Nonverbal communication: The unspoken dialogue* (2nd ed.). New York: McGraw-Hill.
Byron, K. (2008). Carrying too heavy a load? The communication and miscommunication of emotion by email. *Academy of Management Review, 33*(2), 309–327.
Byron, K., & Baldridge, D. (2005). Toward a model of nonverbal cues and emotion in email. In *Academy of management annual meeting proceedings* (pp. B1–B6).
Cascio, W. (1999). Virtual workplaces: Implications for organizational behavior. *Journal of Organizational Behavior, 6*, 1–14.
Chapman, D. S., & Rowe, P. M. (2001). The impact of videoconference technology, interview structure, and interviewer gender on interviewer evaluations in the employment interview: A field experiment. *Journal of Occupational and Organizational Psychology, 74*, 279–298.
Connerley, M. L., & Rynes, S. L. (1997). The influence of recruiter characteristics and organizational recruitment support on perceived recruiter effectiveness: Views from applicants and recruiters. *Human Relations, 50*, 1563–1586.
Danesi, M. (2009). *Dictionary of media and communications*. New York & London: M. E. Sharpe, Inc.
Dansereau, F., Graen, G. B., & Raga, W. (1975). A vertical dyad linkage approach to leadership in formal organizations. *Organizational Behavior and Human Performance, 13*, 46–78.
Derks, D., Bos, A. E. R., & Von Grumbkow, J. (2007). Emoticons and online message interpretation. *Social Science Computer Review, 26*(3), 379–388.
Derks, D., Bos, A. E. R., & Von Grumbkow, J. (2008). Emoticons in computer-mediated communication: Social motives and social context. *Cyberpsychology & Behavior, 11*(1), 99–101.
Dienesch, R. M., & Liden, R. C. (1986). Leader-member exchange model of leadership: A critique and further development. *The Academy of Management Review, 11*(3), 618–634.
Dresner, E., & Herring, S. C. (2010). Functions of the nonverbal in CMC: Emoticons and illocutionary force. *Communication Theory, 20*(3), 249–268.
Eden, D. (1984). Self-fulfilling prophecy as a management tool: Harnessing pygmalion. *Academy of Management Review, 9*(1), 64–73.
Eden, D. (1993). Interpersonal expectations in organizations. In E. D. Blanck (Ed.), *Interpersonal expectations: Theory, research, and applications* (pp. 154–178). Cambridge, UK: Cambridge University Press.
Eden, D., & Shani, A. B. (1982). Pygmalion goes to boot camp: Expectancy, leadership, and trainee performance. *Journal of Applied Psychology, 67*(2), 194–199.
Emerson, R. (2011). *Emoticons at work: unprofessional or necessary evil?*. Huffington Post. Retrieved from, http://www.huffingtonpost.com/2011/10/28/emoticons-work-smiley-business-correspondence_n_1063742.html.
Frith, C. (2009). Role of facial expressions in social interactions. *Philosophical Transactions: Biological Sciences, 346*, 3453–3458.
Fusani, D. S. (1994). 'Extra-class' communication: Frequency, immediacy, self-disclosure, and satisfaction in student-faculty interaction outside the classroom. *Journal of Applied Communication Research, 22*, 232–255.
Ganster, T., Eimler, S. C., & Kramer, N. C. (2012). Same same but different!? The differential influence of smilies and emoticons on person perception. *Cyberpsychology, Behavior and Social Networking, 15*(4), 226–230.

Gerstner, C. R., & Day, D. V. (1997). Meta-analytic review of leader-member exchange theory: Correlates and construct issues. *Journal of Applied Psychology, 82*(6), 827–844.

Graen, G. (1976). Role making processes within complex organizations. In M. D. Dunnette (Ed.), *Handbook of industrial and organizational psychology* (pp. 1201–1245). Chicago: Rand-McNally.

Graen, G. B., & Cashman, J. F. (1975). A role-making model of leadership in formal organizations: A developmental approach. In J. G. Hunt & L. L. Larson (Eds.), *Leadership frontiers* (pp. 143–165). Kent, OH: Kent State University Press.

Graen, G. B., Novak, M. A., & Sommerkamp, P. (1982). The effects of leader-member exchange and job design on productivity and satisfaction: Testing a dual attachment model. *Organizational Behavior and Human Performance, 30,* 109–131.

Graen, G. B., & Scandura, T. (1987). Toward a psychology of dyadic organizing. In B. Staw & L. L. Cummings (Eds.), *Research in organizational behavior* (pp. 175–208). Greenwich, CT: JAI Press.

Graen, G. B., & Uhl-Bien, M. (1991). The transformation of professionals into self-managing and partially self-designing contributions: Toward a theory of leader-making. *Journal of Management Systems, 3*(3), 33–48.

Graen, G. B., & Wakabayashi, M. (1994). Cross-cultural leadership-making: Bridging American and Japanese diversity for team advantage. In H. C. Triandis, M. D. Dunnette, & L. M. Hough (Eds.), *Handbook of industrial and organizational psychology* (vol. 4, pp. 415–446). New York: Consulting Psychologist Press.

Guéguen, N., & De Gail, M.-A. (2003). The effect of smiling on helping behavior: Smiling and good samaritan behavior. *Communication Reports, 16*(2), 133–140.

Hebert, B. G., & Vorauer, J. D. (2003). Seeing through the screen: Is evaluative feedback communicated more effectively in face-to-face or computer-mediated exchanges? *Computers in Human Behavior, 19*(1), 25–38.

Huang, A. H., Yen, D. C., & Zhang, X. (2008). Exploring the potential effects of emoticons. *Information & Management, 45*(7), 466–473.

Huffaker, D., & Calvert, S. (2005). Gender, identity, and language use in teenage blogs. *Journal of Computer-Mediated Communication, 10*(2). http://dx.doi.org/10.1111/j.1083-6101.2005.tb00238.x.

Ilies, R., Curşeu, P. L., Dimotakis, N., & Spitzmuller, M. (2013). Leaders' emotional expressiveness and their behavioural and relational authenticity: Effects on followers. *European Journal of Work and Organizational Psychology, 22*(1), 4–14.

Jack, R. E., Blais, C., Scheepers, C., Schyns, P. G., & Caldara, R. (2009). Cultural confusions show that facial expressions are not universal. *Current Biology, 19*(18), 1543–1548.

Jacobs, T. O. (1971). *Leadership and exchange in formal organizations.* Alexandria, VA: HumRRO.

Jibril, T. A., & Abdullah, M. H. (2013). Relevance of emoticons in computer-mediated communication contexts: An overview. *Asian Social Science, 9*(4), 201–208.

Katsuno, H., & Yano, C. R. (2002). Face to face: Online subjectivity in contemporary Japan. *Asian Studies Review, 26*(2), 205–231.

Katsuno, H., & Yano, C. (2007). Kaomoji and expressivity in a Japanese housewives' chat room. In B. Danet & S. C. Herring (Eds.), *The multilingual Internet: Language, culture, and communication online* (pp. 278–301). New York: Oxford University Press.

Katz, D., & Kahn, R. L. (1978). *The social psychology of organizations* (2nd ed.). New York: Wiley.

Kiesler, S., Siegel, J., & McGuire, T. W. (1984). Social psychological aspects of computer-mediated communication. *American Psychologist, 39,* 1123–1134.

Krohn, F. B. (2004). A generational approach to using emoticons as nonverbal communication. *Journal of Technical Writing and Communication, 34*(4), 321–328.

Kurtzberg, T. R., Naquin, C. E., & Belkin, L. Y. (2005). Electronic performance appraisals: The effects of e-mail communication on peer ratings in actual and simulated environments. *Organizational Behavior and Human Decision Processes, 98*(2), 216–226.

Lea, M., & Spears, R. (1992). Paralanguage and social perception in computer-mediated communication. *Journal of Organizational Computing, 2*, 321–341.
Lee, C. (2003). *How does instant messaging affect interaction between the genders?*. Stanford, CA: The mercury project for instant messaging studies at Stanford University. Retrieved from, http://www.stanford.edu/class/pwr3-25/group2/projects/lee.html.
Liden, R. C., & Graen, G. (1980). Generalizability of the vertical dyad linkage model of leadership. *Academy of Management Journal, 23*, 451–465.
Liden, R. C., Martin, C. L., & Parsons, C. K. (1993). Interviewer and applicant behaviors in employment interviews. *Academy of Management Journal, 36*, 372–386.
Liden, R. C., & Maslyn, J. M. (1998). Multidimensionality of leader-member exchange: An empirical assessment through scale development. *Journal of Management, 24*(1), 43–72.
Liden, R. C., Sparrowe, R. T., & Wayne, S. J. (1997). Leader-member exchange theory: The past and potential for the future. *Research in Personnel and Human Resources Management, 15*, 47–119.
List of emoticons. (n.d.). In *Wikipedia*. Retrieved December 26, 2014 from: http://en.wikipedia.org/wiki/List_of_emoticons.
Lo, S. (2008). The nonverbal communication functions of emoticons in computer-mediated communication. *CyberPsychology & Behavior, 11*(5), 595–597.
Luor, T., Wu, L., Lu, H., & Tao, Y. (2010). The effect of emoticons in simplex and complex task-oriented communication: An empirical study of instant messaging. *Computers in Human Behavior, 26*, 889–895.
Markus, M. L. (1994). Finding a happy medium: Explaining the negative effects of electronic communication on social life at work. *ACM Transactions on Information Systems, 12*, 119–149.
Martin, R., Epitropaki, O., Thomas, G., & Topakas, A. (2010). A review of leader-member exchange research: Future prospects and directions. *International Review of Industrial and Organizational Psychology, 25*, 35–88.
Mayfield, M., & Mayfield, J. (1996). A test of the moderating effects of job autonomy on LMX outcomes: A meta-analysis with performance and job satisfaction. In *Proceedings of the 38th annual meeting of the southwest academy of management* (pp. 251–255).
Mehrabian, A. (1969). Some referents and measures of nonverbal behavior. *Behavioral Research Methods and Instrumentation, 1*, 213–217.
Nguyen, D. T. (2008). *Visually dependent nonverbal cues and video communication*. Unpublished doctoral dissertation, Berkeley: University of California.
Purvanova, R. K., & Bono, J. E. (2009). Transformational leadership in context: Face-to-face and virtual teams. *The Leadership Quarterly, 20*(3), 343–357.
Rezabek, L. L., & Cochenour, J. J. (1998). Visual cues in computer-mediated communication: Supplementing text with emoticons. *Journal of Visual Literacy, 18*, 201–215.
Riggio, R. E. (2005). Business applications of nonverbal communication. In R. S. Feldman & R. E. Riggio (Eds.), *Applications of nonverbal communication* (pp. 119–138). Mahwah, N.J.: L. Erlbaum Associates.
Ruggieri, S. (2009). Leadership in virtual teams: A comparison of transformational and transactional leaders. *Social Behavior & Personality: An International Journal, 37*(8), 1017–1021.
Ryan, R. V. (2009). *Leader-member exchange quality and the role of personality congruence between leaders and subordinates*. Unpublished doctoral thesis, Australia: The University of Western Australia.
Scandura, T. A., Graen, G. B., & Novak, M. A. (1986). When managers decide not to decide autocratically: An investigation of leader-member exchange and decision influence. *Journal of Applied Psychology, 11*, 579–584.
Seaton, D. (2011). *Emoticons, text-speak pop up in emails*. Detroit, MI: The South End. Retrieved from, http://www.thesouthend.wayne.edu/article_bb8f33e3-9a46-56d0-86cf-43b3a7e187b7.html.

Teven, J. J. (2007). Effects of supervisor social influence, nonverbal immediacy, and biological sex on subordinates' perceptions of job satisfaction, liking, and supervisor credibility. *Communication Quarterly, 55*(2), 155–177.

Thompson, L. F., Mullins, A., & Robinson, B. (2010). E-screening: The consequences of using "smileys" when e-mailing prospective employers. In *Presented at the 25th annual conference of the Society for Industrial & Organizational Psychology, Atlanta, GA.*

Uhl-bien, M. (2003). Relationship development as a key ingredient for leadership development. In S. E. Murphy & R. E. Riggio (Eds.), *The future of leadership development* (pp. 129–148). Mahwah, NJ: Lawrence Erlbaum Associates, Inc.

Utz, S. (2000). Social information processing in MUDs: The development of friendships in virtual worlds. *Journal of Online Behavior, 1*(1). Retrieved May 26, 2012 from the World Wide Web, http://www.behavior.net/JOB/v1n1/utz.html.

Walther, J. B. (1992). Interpersonal effects in computer-mediated interaction. *Communication Research, 19*, 52–90.

Walther, J. B. (1996). Computer-mediated communication: Impersonal, interpersonal, and hyperpersonal interaction. *Communication Research, 23*, 3–43.

Walther, J. B., & D'Addario, K. P. (2001). The impacts of emoticons on message interpretation in computer-mediated communication. *Social Science Computer Review, 19*(3), 324–347.

Walther, J. B., Slovacek, C. L., & Tidwell, L. C. (2001). Is a picture worth a thousand words?: Photographic images in long-term and short-term computer-mediated communication. *Communication Research, 28*(1), 105–134.

Wei, A. C. Y. (2012). *Emoticons and the non-verbal communication: With reference to Facebook.* Unpublished master's thesis, Bangalore, India: Department of Media Studies, Christ University.

Wolf, A. (2000). Emotional expression online: Gender differences in emoticon use. *Cyber-Psychology & Behavior, 3*(5), 827–833.

Yuasa, M., Saito, K., & Mukawa, N. (2011). Brain activity when reading sentences and emoticons: An fMRI study of verbal and nonverbal communication. *Electronics and Communications in Japan, 94*(5), 17–24.

Yuki, M., Maddux, W. W., & Masuda, T. (2005). Are the windows to the soul the same in the East and West? Cultural differences in using the eyes and mouth as cues to recognize emotions in Japan and the United States. *Journal of Experimental Social Psychology, 43* (2007), 303–311.

CHAPTER 4

Empathetic Technology

Néna Roa Seïler[a], Paul Craig[b]
[a]Edinburgh Napier University, Edinburgh, UK
[b]Xi'an Jiaotong-Liverpool University, Suzhou, China

INTRODUCTION

To start looking at empathy and technology, we begin by looking at the philosophical and psychological underpinnings of empathy. Next we consider the relationship between empathy and technology, and look at some existing models of empathy related to technology. We continue by considering the concept of the virtual agent as an empathetic companion and describe some examples of virtual companions in commercial and academic settings. To conclude, we look at the results of user studies focused on understanding our perception of virtual companions and describe what is required to develop an appropriate companion interaction strategy.

The chapter will give the reader grounding in the essential theory of empathy and its relationship with technology. It will also introduce state-of-the-art empathetic agents and describe how experimental methods such as the use of the Kelly grid and Semantic Differential can be used to discover how users feel about emergent technologies. The chapter will be of interest to application designers or developers working with interfaces that employ agents or have agent-like behavior, or anyone who wants to find out more about the developing relationship between empathy and technology.

What is Empathy?

Empathy is the ability to understand and respond to the unique affective experiences of another person (Decety & Jackson, 2006). If we consider empathy at the level of experience, it is essentially a psychological construct that denotes a sense of similarity between one's own feelings and those expressed by another person. Empathy can be also seen as an interaction between two individuals who share each other's experiences and feelings, although this exchange of feelings does not necessarily mean that one will act or even feel compelled to act in support or sympathize. Indeed, the social and emotional situations that arise due to empathy can be quite complex.

These depend on the feelings experienced by the observed person, the relation of the target to the observer, and the context in which the social interaction occurs.

Since the age of classical antiquity, empathy has been seen as important because it allows us to be able to understand the thoughts of others and to predict or explain what others feel, think, or do. It has also been seen as having a central role in ethical thinking and behavior. The "Golden Rule" of ethical reciprocity (Wattles, 1997) relies fundamentally on the psychological process of empathy. This is often cited as a pillar of the Christian faith (Armstrong, 1994), classically stated in the Bible as "Do unto others as you would have them do unto you," and is prominent in other religious traditions such as Hinduism, Buddhism, Taoism, and Zoroastrianism. Empathy is also considered to have a central role in the esthetic understanding of our engagement with works of art and fictional characters. Following on from these philosophical and religious underpinnings, a more complete understanding of empathy is now offered by the interaction of research in science and humanities (Coplan & Goldie, 2011).

Recently, empathy has received much attention from philosophers, psychologists, and cognitive neuroscientists. Here, studies have documented that empathy plays a central role in moral reasoning and pro-social behavior that motivates and inhibits aggression toward others. Batson, Duncan, Ackerman, Buckley, and Birch (1981) are part of this movement, offering an empathy-altruism hypothesis which states that reliable, purely altruistic action can only happen if it is preceded by empathic concern for another. Other authors define empathic concern as an emotional reaction characterized by feelings such as compassion, tenderness, sympathy, and soft heartedness (Decety & Lamm, 2006). Conversely, a lack of empathy is seen as leading to aggressive, antisocial behavior (Miller & Eisenberg, 1988) and cruelty (Baron-Cohen, 2011). Other authors consider empathy as socially oriented emotion, defined as "the ability to put oneself into the mental shoes of another" (Goldman, 1993), a complex form of psychological inference (Ickes, 1997), an affective response more appropriate to someone *else's* situation (Hoffman, 1982), another-oriented emotional response (Batson et al., 1997), or an affective response that stems from the apprehension or comprehension of another's emotional state (Eisenberg, 2000). In the field of cognitive neuroscience, there is a further distinction between cognitive empathy which is the ability to *know* what another person is thinking and feeling, and affective empathy which is the ability to actually *feel* another person's emotional state (Rueckert & Naybar, 2008).

Bringing these different definitions together, empathy is essentially the ability to imagine or *feel* ourselves in the position of others, and it is generally expected that this insight will guide us toward more ethical or moral behavior. Empathy allows us to perceive the experience and feelings of others as if those experiences and feelings were shared, and encourages us to improve the situation of others as if it were our own. These feelings of empathy can also help explain altruistic and unselfish behavior where actions are exclusively for the benefit of others. Empathy also allows us to be more tolerant and accepting of others. This is particularly important in this modern age, as working and learning environments become increasingly diverse due to high levels of global migration (Hollingsworth, Didelot, & Smith, 2003; Joplin & Daus, 1997). Indeed, common empathy is often cited as an important factor for cohesion and maintenance of social structures within a healthy society (Bartal, Decety, & Mason, 2011; Byrne et al., 2008; Pierce, 2008).

EMPATHY AND TECHNOLOGY

At first glance, empathy and technology do not seem to be compatible. Empathy is considered to be a very human skill. It involves subtle spoken and visual cues that are often transmitted subconsciously and can be difficult to read or mimic. Empathy also involves human attributes such as kindness and compassion that we do not associate with computers. Indeed, computers are often considered the antithesis of empathetic thinking and this is reflected in our language. If a person lacks empathy we might describe them as a *robot* or a *machine*. When someone lacks empathy they are *not human* because empathy is a big part of what we think of as being human. On the other hand, if a computer is *emotional*, this is seen as a bad thing too. If a computer does not work well we might say it is *temperamental* or it *knows what we're thinking*. This probably means it has stopped working just at the point when we want to do something important. We also know that computers process data faster and can perform calculations at a rate incomprehensible to humans, so it is natural that we should be somewhat apprehensive of computers developing our most *human* abilities, lest they supersede us in our position of authority. This nightmare is a common theme of modern cinema, reappearing in productions such as *2001: A Space Odyssey* (Kubrick & Kubrick, 1968), *The Terminator* (Hurd, & Cameron, 1984), *The Matrix* (Silver, Wachowski, & Wachowski, 1999), and, more recently *Her* (Ellison & Spiegel, 2013), a film about a man who falls in love with his emotionally active operating system, only to be jilted when the machine realizes that it/she has outgrown him intellectually.

There are, however, some serious disadvantages to computers not being emotional and empathetic entities. We are starting to spend an increasing proportion of our time with computers and a lot of this time we are using computers to interface with other human beings. If our interaction with these machines is unemotional and without empathy there is a real danger that we might begin to communicate with other human beings the same way we communicate with our computers (Nishida, 2013). If we are communicating *through* a machine that is incapable of supporting empathy, there is a danger that we be less compassionate when talking to people too. Human manifestations of these problems are "the social retardation of computer dependent youth" (Fomichov & Fomichova, 2014) and online harassment or cyber-bullying (Heirman & Walrave, 2008; Hinduja & Patchin, 2010; Slonje & Smith, 2008; Smith et al., 2008). These have become significant problems facing modern society.

Other examples of technology seeming to reduce empathy are: nurses becoming less empathetic when using increasing amounts of technology in intensive care units (Brunt, 1985), technology reducing the level of service in customer care environments (Gorry & Westbrook, 2011), and students becoming dehumanized when teachers rely excessively on computers (Nissenbaum & Walker, 1998). Moreover, the need to make technology more empathetic and to recognize users as complete humans beings has been accepted for some time (Muller, Wharton, McIver, & Laux, 1997), and yet, the Human Computer Interaction community has continued to focus almost exclusively on aspects of human behavior related to efficiency and productivity (Picard & Klein, 2002).

The implementation of empathetic interfaces present massive technical challenges relative to simulating aspects of empathy, the difficulty in modeling empathy, and the sea-change in attitudes toward computers that would be needed in order for people to accept empathic machines.

The magnitude of the problem depends on the level of empathy to be incorporated into an interface. At the basic level, if we simply decide to consider an empathetic interface design by paying more attention to the user's *feelings* toward a normal product (Crossley, 2003), there are no obvious additional technical challenges. On the other end of the scale, deciding to emulate the full gamut of human emotional empathy would be an enormous technical undertaking. Real human empathy relies on the transmission of perception of a range of subtle visual and audible cues that work with varying degrees of conscious awareness on the part of either transmitter or observer. The mechanisms involved in processing emotional information and formulating appropriate responses are also very far from being well understood. Indeed, it is questionable whether these processes can even be fully understood given the massive complexity of human processing, the variety of factors involved in assessing a social situation (McQuiggan & Lester, 2007), and the level of variation among cultural groups and individuals (Davis, 1980). It is evident that a computer emulating human empathy could only include parts of real human empathy. Hence, when we try to model human empathy it is important to consider (a) what to include for a model to be effective and (b) how to model that part of human empathy. The ultimate test for any such model would be for the user to *accept* the empathy in the context of using a machine, which is normally thought of as cold, unthinking, and without feeling. This is further complicated by phenomena such as the "uncanny valley effect" (Mori, 1970) which holds that when a machine has a level of near-human appearance it can have an unsettling effect on the observer. This could easily extend to near-human *behavior* since our behavior often affects how we appear.

MODELING EMPATHY

The two principal approaches to modeling empathy for technological applications are the analytical and the empirical methods (McQuiggan & Lester, 2007). In the analytic approach, models are constructed by analyzing literature dealing with empathy. The empirical approach uses observations of empathetic behavior.

The fundamental problem with the analytic approach is that empathy is still not well understood and has only become an established field of study for psychologists within the past 25 years (Davis, 1994). This is compounded by the innate complexity of empathy, which involves the comprehension of another person's intentions, affective state, and situational context. This means that devising a *universal* model of empathy is likely to be beyond practical considerations for the foreseeable future (McQuiggan & Lester, 2007).

Indeed the computational modeling of human emotions is seen as one of the big challenges of Artificial Intelligence (AI). Existing models can identify, model, and imitate human emotions to a limited degree by using theories of psychology (Marsella & Gratch, 2014) and more recently computational models of empathy have started to appear based on the theory of appraisal, also known as the theory of cognitive evaluation of emotions.

AI models that permit empathetic responses include those of Rodrigues, Mascarenhas, Dias, and Paiva (2015) and Boukricha, Wachsmuth, Carminati, and Knoeferle (2013). Both researches investigate the factors that modulate the emphatic behavior of agents. This allows them to display different degrees of empathy. The intensity of empathy during the interaction is controlled by different features such as similarity, affective link, mood, and personality. The model was tested in a bullying scenario to measure if agents with the ability to empathize could moderate between bully and victim. The agents used had internal processes based on psychological models to mimic human empathic behavior and results suggest that when agents have similar emphatic responses, the ones with empathetic behavior are perceived as more caring, likeable, trustworthy, intelligent, dominant, and less submissive.

There is, however, no straightforward way to test a computational model like this one in isolation from a concrete application and scenario (Aylett & Paiva, 2012). This is a serious limitation for the testing of emphatic agents, especially when an end user is interacting in real time and it is important to measure the global user experience.

Empirical approaches using observations of real empathic behavior show more promise of generating what we might consider as *less-than-complete* yet useful models of empathy. These can be targeted to suitably extend the communicative capabilities of socially intelligent agents rather than to the generation of universal models. Indeed, they should be particularly useful when used in scenarios involving end users interacting in real time. Examples of such methodologies are CARE, a data-driven affective methodology for learning empathy models, and the star-model, an empirical approach adjusted for the interaction of virtual companions (Benyon & Mival, 2007; Smith et al., 2011). Other models are based on the strategies used by human tutors to keep their students engaged. These include the models of Woolf and Burleson (Burleson Daviss et al., 2006; Woolf et al., 2009) which sense the learner's level of frustration and assess the appropriate task based intervention for a virtual companion.

In general, empathetic computational systems have four main components. These are: perception, modeling of the user's emotional state, processing, and feedback. Perception of an emotional state can be through any combination of several channels using either semantic information, interpretation of visual or audible cues, or even recording skin resistance or heart-rate (Prendinger & Ishizuka, 2005). Once this information is processed, the user's emotional state is characterized by using a model of emotions such as the Russell model (Russell, 1980), based on two dimensions of arousal and valence (see Figure 4.1), or the Plutchik wheel (Plutchik, 2001), which includes pure and overlapping versions of different degrees of the eight

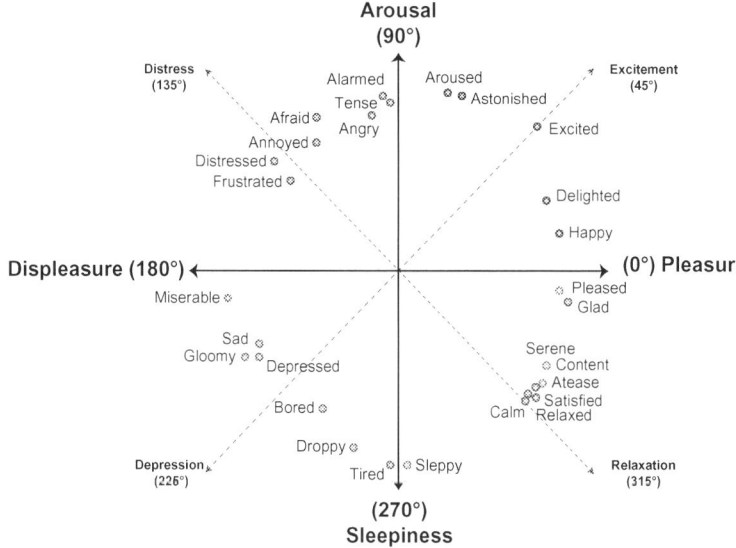

Figure 4.1 The Russell dimensional model for emotions (Russell, 1980).

primary emotions (see Figures 4.2 and 4.3). The next stage is processing to combine emotional information with semantic content to formulate appropriate feedback.

Other, more sophisticated models (Roa-Seïler, Benyon, & Leplâtre, 2009) use something known as the *backchannel* to deliver a more timely, less

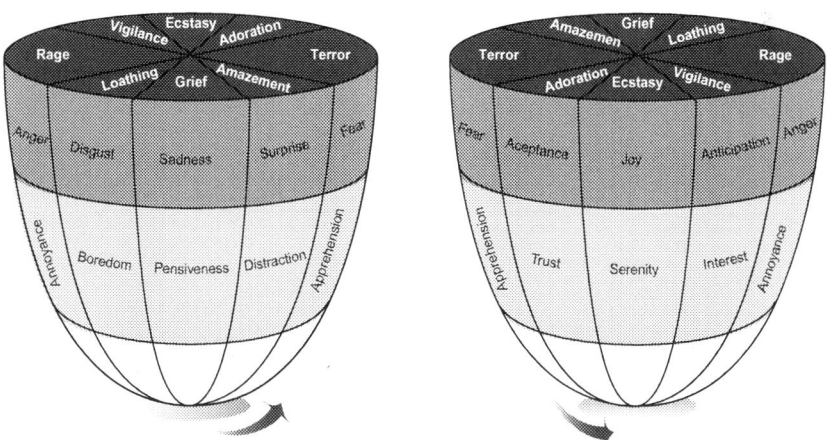

Figure 4.2 Plutchik's wheel for modeling emotions (Plutchik, 2001). *Courtesy of Dr. Néna Roa Seïler, unpublished doctoral dissertation.*

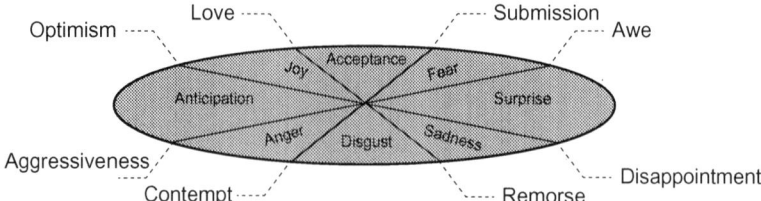

Figure 4.3 Description of mixed emotions (Plutchik, 2001). *Courtesy of Dr. Néna Roa Seïler, unpublished doctoral dissertation.*

accurate immediate response, prior to a more reasoned, well thought-out response. Backchannel responses are more likely to be interjections or short affirmations that communicate emotional empathy rather than semantic content. This confirms the fact that more measured responses take more time to deliver because of the complications involved in properly interpreting stimulus and formulating appropriate semantic responses. This model of backchannel responses also corresponds with known neurobiological processes (LeDoux, 1995) where emotional signals pass simultaneously through the faster Amygdala and the slower but more accurate Sensory Cortex to express initial and later emotional responses (see Figure 4.4).

An initial Amygdala response to an animal noise at night might be a yelp or a jump with the later Sensory Cortex response being to ask what animal made the noise or remark as to the quality of the sound. An Amygdala

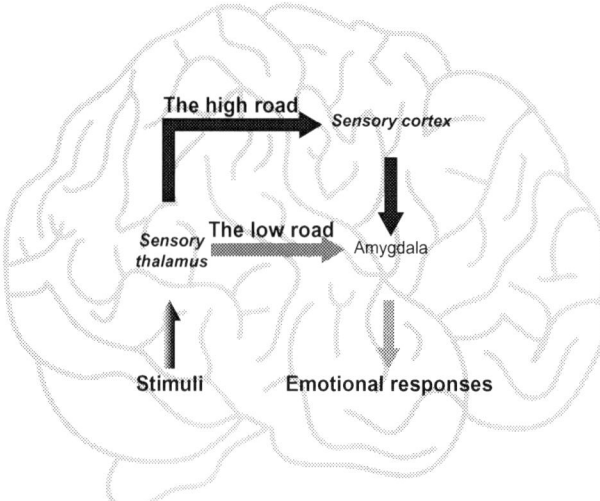

Figure 4.4 Neurological structure of emotions according to LeDoux (1995). *Courtesy of Dr. Néna Roa Seïler, unpublished doctoral dissertation.*

response to a crude joke might be a guttural laugh while the later Sensory Cortex response could be shame or revulsion.

EMPATHETIC VIRTUAL COMPANIONS

Empathetic Virtual Companions are virtual, screen-based characters that form an empathetic emotional connection with the user over time (Benyon, Hansen, & Webb, 2008; Benyon & Mival, 2007; Roa Seïler & Benyon, 2010). Companions introduce a significant change in the relationship between people and technologies because companionship is about an accessible, pleasing relationship with an interactive source in which there is social and emotional investment. Companions also require a level of trust, compatibility, and familiarity within this relationship that results in a feeling of security, contentment, and general well-being (Benyon & Mival, 2013). This is significantly different from our established idea of technology where communication is predominantly unidirectional and impersonal.

In more technical terms, Companions are considered no-task-oriented Embodied Conversational Agents and are essentially the evolution of Intelligent Tutoring Systems that assume a dual role as both intelligent tutor and learning companion (Chou, Chan, & Lin, 2003). Studies have already demonstrated that these learning tutors are almost as effective as individual tutoring by experienced trainers with the benefit that they require a lot less instructor time (Reif & Wang, 1999). They can also add significant value to existing learning environments that do not normally require an instructor (Craig, Roa-Seïler, Díaz, & Rosano, 2013). Other task-oriented agents with companion-like behavior are used to form social relationships with users for higher motivation (Bickmore, 2003), reduce stress in job-interview role play sessions (Prendinger & Ishizuka, 2005), teach children how to deal with frustration in the classroom (Burleson & Picard, 2007), help children deal with bullying (Paiva et al., 2005), and act as health coaches (Bickmore & Cassell, 2000). All of these activities rely heavily on the agent's ability to develop a relationship with the user and demonstrate empathetic behavior.

The defining feature of Empathetic Virtual Companions is that they evoke feelings of companionship in the user. Companionship can however be difficult to define. At a literal level, it is a feeling of fellowship or friendship (Simpson & Weiner, 1989), but these concepts themselves can be difficult to define and it is not clear whether they encapsulate all that we mean by companionship. For many, companionship implies a more long-term relationship with a degree of empathy and intimacy. Other definitions of companionship

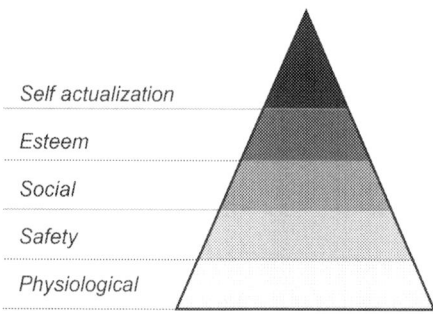

Figure 4.5 Maslow's hierarchy of social needs (Maslow, 1943). Companionship belongs at the level of Social needs.

come from the science of psychology, which considers companionship to be part of a hierarchy of social needs (Maslow, 1943), defined as a social need along with love and belonging. This positions companionship as higher than the need for security and lower than the need for esteem (see Figure 4.5). Other authors focus more on the mutuality of companionship, defining it as shared caring and trust (Gleitman & Papafragou, 2005). These definitions tend to involve reciprocal behavior and empathy is necessarily a big part of this.

In the context of technology, a key part of the acceptance of Companions by users is the emotional aspect of companionship. It is also suggested that in order for a system to function successfully as a Companion, it must be able to appear human and pass the Turing test based on its ability to mimic human conversation (Cowie, 2010). On the other hand, considering the relationship that humans can develop with pets such as dogs and cats, other authors suggest that neither the ability to hold a conversation nor a human appearance are necessary elements in a Companion. A Companion simply needs to recognize their owner as an individual, have good intentions toward him, display predictable behavior, be able to predict their owner's behavior and appear independent (Pulman, Boye, Cavazza, Smith, & De La Cámara, 2010).

Whether a complete technological implementation of what these authors define as a companionship is desirable, or even possible, is debatable. For example, it is impossible to know if a machine can *recognize their owner as an individual* without first tackling some quite contentious philosophical issues. There are, however, a number of applications already developed as virtual Companions and these have already succeeded in giving the impression of Companion-like behavior with some degree of success. Hence, the ability to develop virtual companionship cannot be considered beyond our human capacity or outside the constraints of current technology.

One of the earliest incarnations of a virtual agent with Companion-like behavior is the Microsoft Office Virtual Assistant. This commonly took the form of an animated cartoon paper-clip called Clippy (see Figure 4.6) that would popup with help or suggestions when the user decided to activate certain functions of the software. Clippy was, however, incredibly unpopular with users who found him annoying and above all, frustrating (Shneiderman & Plaisant, 2005). Clippy was inappropriately intrusive and often required unnecessary interaction at inopportune times. Part of this problem can be attributed to its lack of empathy. Clippy would often popup without being needed. If he were able to empathize or sense that the user was happy using the software at any given time, he could have adapted to be less intrusive. There was also a sense that Clippy was oblivious, and therefore uninterested, in the user's emotional state. Clippy was constantly smiling with wide eyes and an open expression. He appeared eternally exuberant and when the user was less than happy, this expression could be perceived as callous or mocking. A more empathetic response would be emotionally neutral or sympathetic.

Clippy was almost universally hated (Deegan, 1999; Luehning, 2001), and perhaps the most important lesson to be taken from this is that whenever Companion interfaces fail, they fail spectacularly. When we are faced with an interface that appears human, or has human features like Clippy, we expect human levels of emotional intelligence. Subsequently, people feel more let down when faced with a human-like agent that does not function

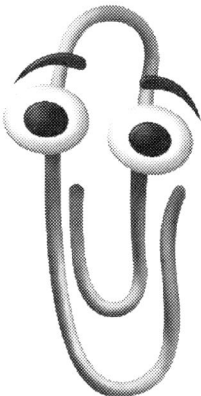

Figure 4.6 Clippy, the Microsoft Office virtual assistant.

as expected (Vugt, Bailenson, Hoorn, & Konijn, 2010). Despite this, Microsoft persisted with Clippy, including it in all versions of Office between 1998 and 2004, probably recognizing the potential for Companion interfaces without quite being able to develop more user friendly implementation.

The Sony Enterprise developed AIBO in a somewhat more successful attempt at a virtual Companion (Artificial Intelligence Robot or Aibo, the Japanese word for pal or partner, see Figure 4.7). These are robotic dog-like pets with artificial vision, facial, and speech recognition that can respond to commands and use an LED system to express emotions such as happiness, anger, sadness, and fear. Results show that people find it easy to attribute real emotions and intelligence to AIBO (Benyon & Mival, 2013) and AIBO has even been used as a Companion to combat loneliness in animal-assisted therapy projects (Banks, Willoughby, & Banks, 2008).

Other virtual pets similar to AIBO are Furbies and Tamagochi. Furbies resemble something between a hamster and an owl. The original Furbies, launched in 1998, began by speaking their own native language and gradually, over time, introduced words from English (or the user's language, depending where the toy was sold) into their random vocabulary to give the illusion that they were learning. The latest Furbies include voice recognition and limited emotional responses. Tamagochi are entirely screen-based entities who require the user to service its basic needs for food, water,

Figure 4.7 AIBO, the robot dog virtual Companion developed by Sony.

love, etc. regularly, in order for it to grow, develop, and maintain its emotional state. If the Tamagochi becomes too sad or hungry, it dies.

Pet Companions such as AIBO, Furbies, and Tamagochi are generally very popular products that show evidence of the development of companionship-like, intimate emotional relationships with their users (Banks et al., 2008; Benyon & Mival, 2013; Luh, Li, & Kao, 2015). This is perhaps because appearing as an animal or pet lowers the user's expectations to less than what they might expect from a human Companion. These Companions also avoid the complexities of recognizing and decoding the semantic content of speech. This demonstrates that feelings of companionship are possible with a limited range of what we consider constitute companionship in a human relationship.

Another important application area for virtual Companions is driving. People are already familiar with the use of humanized speaking technology while driving with intelligent GPS systems that provide incremental real-time directions. The integration of emotionally aware GPS systems would seem like a natural progression, since driving can be a stressful and emotionally engaging activity. In-car driving Companions include AIDA (Williams, Peters, & Breazeal, 2013) and the operating system of the Nissan Pivo concept car. Compared to traditional GPS, these systems are found to reduce stress and encourage safer driving while developing a feeling of companionship with their users. The difference between current implementations of driving-companions and our classic conceptualization of a virtual Companion is that driving-companions are specifically task-oriented.

A more complete implementation of a virtual Companion is offered by Samuela (Roa-Seïler et al., 2014) (see Figure 4.8), a screen-based avatar

Figure 4.8 Samuela the virtual Companion. *Courtesy of Dr. Néna Roa Seïler, unpublished doctoral dissertation.*

developed by the European Union funded Companions project. Samuela is designed to be part of the *home of the future,* living with her owner providing good company, utilizing a variety of different strategies to deliver comfort and encouragement. Currently, Samuela provides an interface for two services developed by the Spanish telecoms company Telefonica. The Cooking Companion is an application that provides users with cooking and dietary advice. It allows people to choose from a selection of dishes, making their choices either through touch or speech. Annota is a tool to manage shared text, video, or phone messages. Later implementations of Samuela are expected to recognize users and provide specialized help based on knowledge of habits and preferences learned from prior interactions.

PERCEPTIONS OF EMPATHETIC VIRTUAL AGENTS

We already know that virtual Companions work well as non-humanoid pets or in task-oriented scenarios where they are endowed with a limited range of what we might consider as full human companionship attributes. This section looks at people's perception of Companions to gain a better understanding of what needs to be implemented in a virtual Companion for it to be accepted and useful. Important questions include—how do people perceive ECAs acting as Companions? What are users' feelings toward robots, screen-based avatars, communicating objects, and toward media using emergent technologies such as speech and visual recognition? How do people feel about computers with anthropomorphic features, and what sort of mental images are evoked by these artifacts? Speaking and conversing are also important for human perception. When interacting with conversational agents, we need to know if people really consider them as conversational agents or simply as speaking agents. Another important question is whether or not people expect a Companion's behavior to be as human or do they simply expect them to exhibit behavior that is consistent with human behavior? These are *all* important questions to be answered in order to design virtual Companions that properly serve human users' needs.

There are numerous theories examining how people interact with objects and in the last few years enormous interest has been shown in studying feelings triggered by objects and artifacts (Benyon & Mival, 2013; Norman, 2002; Picard & Klein, 2002; Roa Seïler & Benyon, 2010). Taking the semiotic philosophy of Charles Sanders Pierce as a frame of reference, Vihma (2003) points out that interaction with objects is not reduced just to the physical or symbolic level, but relies on both. Using the same approach, Rohlin (2002) argued that perception of the usability attributes of products is an unconscious process and it is therefore difficult to find a link between esthetics and the usability of products.

Sense-making is a process of understanding the meaning of objects to people. This involves concepts being expressed by words with two opposite meanings, a denotative meaning and a connotative meaning. Denotative meaning is the semantic meaning of the word. The connotative meaning is the affective meaning, the symbolic added value provided by the speaker (Barthes, 1965). These are used to measure the meaning of concepts, using a bipolar scale known as the Semantic Differential (Osgood, Suci, & Tannenbaum, 1957) (see Figure 4.9). This does not provide information about the meaning of the concept but rather about the emotion, feelings, and therefore the attitudes that the word elicits.

The Personal Construct Theory (PCT), developed by George Kelly in 1950 (Adams-Webber & Kelly, 1979), proposes an additional focus, one on the comprehension of mental meanings. According to Kelly's theory, scientists and users alike are observers of the world around them. Like scientists, they draw up hypotheses which they check through their life experiences to elaborate their own theories and construct their particular perspective (Neimeyer & Bridges, 2004). In other words, individuals have concepts or references called constructs, which allow them to make sense of the world. These constructs also help a person assess their environment and this process has an important impact on their decisions (Kelly, 2003).

The repertory grid, also known as the Kelly grid, is a technique for uncovering people's concepts and the values they call on to understand things. It indicates which concepts people are attached to and what the influences of these are, thus revealing how their mental constructs operate (Fransella, Bell, & Bannister, 2004; Kelly, 2003). This flexible tool is used in fields such as architecture (Lawson, 2007), product design (Hjort af Ornäs, 2010), interactive system design (Tan & Tung, 2003), and software development (Maller et al., 2012). It can also be used to reveal users' perceptions of emerging technologies, facilitating more informed development of new interfaces.

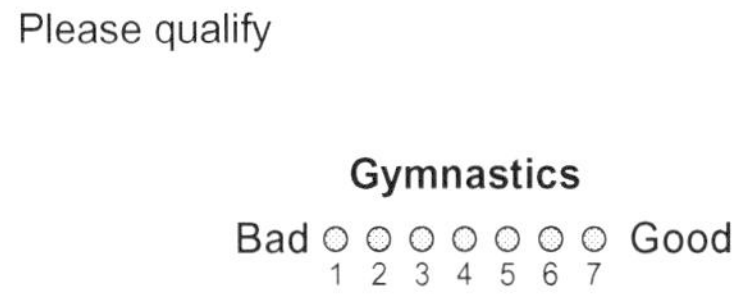

Please qualify

Gymnastics
Bad ○ ○ ○ ○ ○ ○ ○ Good
1 2 3 4 5 6 7

Figure 4.9 Semantic Differential to elicit feelings toward the word Gymnastics.

In order to improve our understanding of how best to develop future Companion interfaces, we undertook a number of experiments to reveal users' global perceptions of Companions and companionship. We began by classifying a set of nine representative Companions relative to two dimensions: materiality and engagement. Here, materiality relates to the embodiment of the proposed Companion, and engagement to the feelings that the Companion elicited during interaction with users. In addition to this, we looked at the references people used to evaluate functions and the impact on the relationship between the embodiment of the Companions and their functions. The goal of the exploratory work was to illustrate these concepts by looking into the meanings attributed to a set of Companions; discussing the repercussions of these meanings on studies of users and on the future of designing Companions.

In order to prepare our Kelly grid, nine representative Companions were positioned into the corners of what is known as the Semiotic Square (see Figure 4.10). The semiotic square is a tool that allows the formalization of the binary relationship between contrary semiotic signs. This is considered to be the elementary structure of meaning (Greimas, 1968). In Figure 4.10, S_1 and S_2 are presented as opposite concepts, S_1 and $\sim S_1$, S_2 and $\sim S_2$ are contradictory concepts, whilst S_1 and $\sim S_2$, S_2 and $\sim S_1$ are complementary concepts.

For the semiotic square in our experiments, we used nine representative virtual Companions from research, commercial products, and science fiction books. These were specially chosen because of their different levels of embodiment and engagement, and included:
- Nabaztag—A Wi-Fi enabled ambient electronic device in the shape of a rabbit

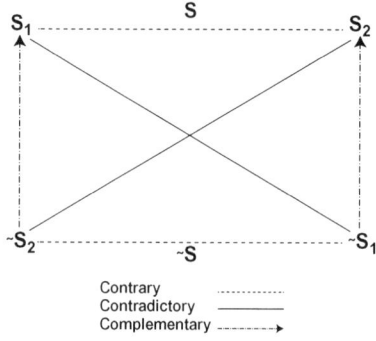

Figure 4.10 The Semiotic Square (Greimas, 1968).

- Aibo—Robotic virtual pet
- Ifboot—Toy robot
- Samuela—Companion avatar
- Patachon—Screen-based virtual pet
- Enrica—Task-oriented screen-based Embodied Conversational Agent
- Pivo—In-car driving computer
- Nike with iPod—Program for running that receives GBS data from the user's training shoe
- Hal—The sentient computer in the film "2001 A Space Odyssey"

Figure 4.11 shows how these Companions are positioned according to materiality and affect. The Companions to the right with higher materiality all have a physical form, while those on the left do not. The Companions in the upper half are all able to demonstrate affect toward a user and therefore they could develop an engaging relationship with users.

Our Kelly grid experiments took place with 15 users of different sexes, age groups, and levels of computer literacy. Each participant was shown videos of each Companion in action then shown five different sets of three cards, each with one of the nine Companions. For each set of cards they

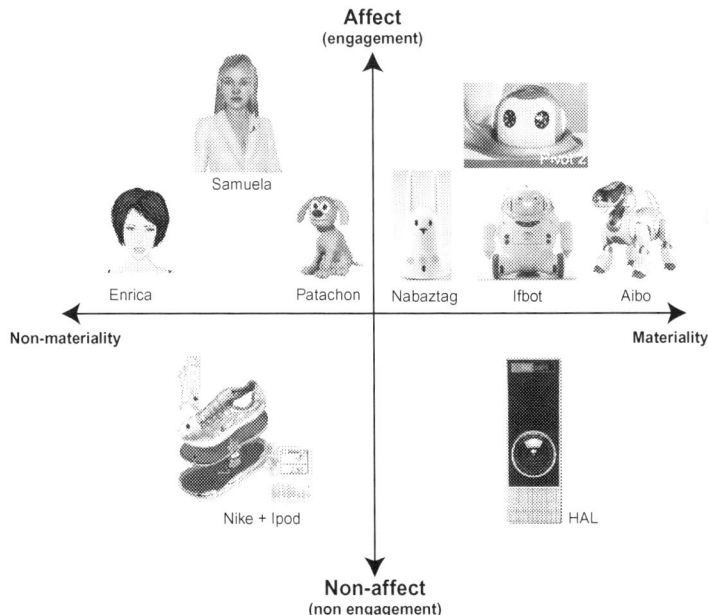

Figure 4.11 Affect and materiality for nine representative Companions. *Courtesy of Dr. Néna Roa Seïler, unpublished doctoral dissertation.*

Please choose three Companions from the previous grid:

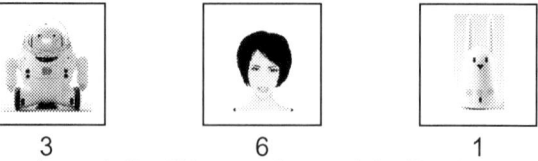

Now, compare them to decide which two are the most similar. Write down why you think the two are similar and why the other one is different

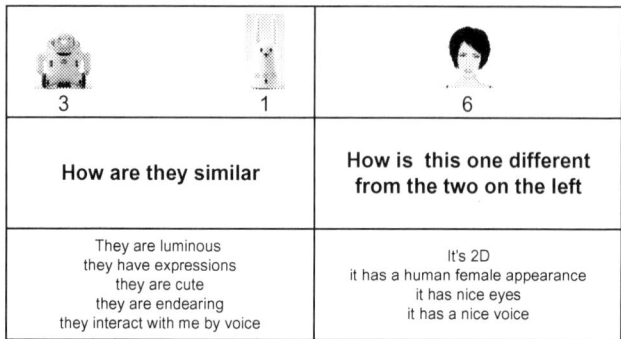

Figure 4.12 Example of Pattern Grid for experiment. *Courtesy of Dr. Néna Roa Seïler, unpublished doctoral dissertation.*

were asked how each of the two most similar cards where similar and how each third card was different from the first two. Figure 4.12 shows an example response for a single set of three cards.

The results of the grids were gathered together and used in a focus group to describe how users felt about each Companion and Companions in general. The adjectives assigned to each Companion were:
- Nabaztag: Toy, uninteresting, colorful, light, radio substitute, luminous, helps searching without tiredness, robotic, like an animal, for children, kindly, cute, makes signs with body.
- Aibo: Toy, dog, dog with expressions, cute, funny, silent, follows instructions, facial light, luminous, time waster, virtual answer, animal, for children, submissive, imitates a living creature, expressive, expression in the eyes, can be tamed, makes signs with body, expresses emotions.
- Ifboot: Follows instructions, light, facial expression, silent, rounded head, soft colors, easy communication, time waster, virtual answer, like an animal, for children, kindly, has body movements, has expressions.
- Samuela: Female face and style, female-eyed, pretty character, pretty face, sympathetic listener, female voice, voice-interactive, can be questioned, useful, like a female speaker, pretty and young, sexy, helpful in

everyday life, interesting, asks and answers questions, virtual look, expresses emotions, acts like a Companion, provides advice on everyday life, has interactions.
- Patachon: Cute, funny dog with expressions, silent, funny, useless, expresses emotions, acts like a Companion, light, has body movements, has expressions.
- Enrica: Female face, pretty face, female-eyed, sympathetic listener, female voice, helpful in everyday life, interesting, asks and answers questions, like a female speaker, pretty and young, sexy, imitates living creatures, expressive, human voice, no real dialog, intimidating because there is no interaction, provides advice on everyday life.
- Pivo: Obeys orders, light, facial expression, rounded head, soft colors, easy communication, helps search without tiredness, robotic, mechanical, related to another machine, nonanthropomorphic expression, like a speaking toy, more virtual, personal use.
- Nikes with iPod: Pretty shoes, for sport minded people, efficient, without dialog, user could decide to use or not, nonhuman-like.
- Hal: Depressing, ugly looking, human voice, no real dialog, intimidating because there is no interaction, virtual, speaking machine, spy camera, frightening, nonhuman.

The analysis of participants' responses provided an insight into the societal impact, that is, the new relationships people might want to develop with Companions as a new interface involving emergent technology. People described their social interaction with Companions and drew a singular approach to how the "Companion hierarchy" could work (my Companion, your Companion, the survey Companion that belongs to a company and so on), illustrating their expectations of the *technological promises* in which the future becomes an object of desire. They also described the level of technology and multimodal exchange they wanted with Companions. As there is no proper semantic for these types of artifacts, users tended to use a variety of metaphors to describe these entities.

It is clear that the arrival of Agents endowed with new human-like abilities such as speech and vision, appearing in different embodiments such as robots, screen avatars communicating *things,* is changing the way people think about objects. Some authors have already named these changes as "Personification Technologies" which predispose people to anthropomorphize (Benyon & Mival, 2008). Nevertheless, mere anthropomorphic attributes seem to be insufficient for transforming an object into a Companion in the same way that simple body movements and facial expressions alone do not. As a result, the Companion Nabaztag is considered to be uninteresting, Companion Ifboot is seen as a time-waster, Companion Hal is considered depressing in spite of his mellow human voice, and the Nike iPod app is only seen, as expressed by test subjects, as *technologically pretty shoes.*

Unusual modes of interaction were also found to create disorientation and confusion in test participants. Enrica is a good example of this; she is able to address people through speech while users answer by typing and this interaction in two different modes results in users considering this companion *intimidating.* Clearly people consider voice to be the effective means of interaction with these characters.

AIBO is another interesting example of this. Because his features allow him to display emotions, he is very successful eliciting a feeling of attachment from his owner. A number of interviews, however, showed that the same owners considered AIBO more like a toy than a Companion because they did not know what he was to be used for.

Indeed, the utility of Companions seems to have particular importance. Without it Companions are not interesting to live with. Users are ready to make concessions with technology if they think the service offered by the Companion is worth it. This usefulness quality has already been explained as "instrumental support" necessary to establish and maintain a relationship (Bickmore & Picard, 2005). Users expect a high degree of usefulness from Companions based on their being perceived as dual entities, as machines, and as humans.

A Companion's functionality and utility are essential to its acceptance by its user. AIBO, although accepted as a good Companion by being capable of interaction with his owner, lacking any usefulness, any task related capacity, it was not, for that reason seen as *good enough,* as users preference is for a good Companion that is also useful.

A Companion is not considered to be *good* if it does not have any sort of a job to do or if it makes no contribution. This instrumentality of involvement is present with nontechnological companions such as humans, pets, and it is this feature that is sought when interacting with tech-companions. It is well known that pets given to children contribute to their social adaptation skills (McNicholas & Collis, 2001). Some authors also consider marriage to be a similar kind of nontechnological companionship which resulted from the social changes of industrialization and urbanization (Edgell, 1972). Studies measuring marital satisfaction confirm that instrumental attitudes are better understood from companionate attitudes (Sillars, Weisberg, Burggraf, & Zietlow, 1990).

Users also expect Companions to have a combination of human-like and agent behavior. This means that people found themselves conjuring up, inventing relationships with agents which would be somewhere between human and object relationship.

It is a striking paradox that users are aware of the fact that Companions are not humans; nevertheless, they expect them to provide the illusion that they are. Thus the paradigm of *computers as social actors* is confirmed (Reeves & Nass, 1996). This goes even further in the case of Companions; it is not only the device itself that needs to pretend to be human, but also the Companion within the device, and users are happy to let themselves be fooled.

To fully experience a digital experience the same way as when interacting with a human being, people need not only a character which provides an illusion of life as claimed by Bates (1994), they also feel compelled to experience a "willing suspension of disbelief," a neuro-psychoanalytic experience in which we humans accept what is represented as though it were really happening (Holland, 2003).

The *human side* of Companions people expect to see seems to be similar to human strategies for capturing audience involvement, such as interaction, humor, and contextualization. Other elements of human strategies for communication and involvement, such as body language, stance, facial expressions, use of space, and gesticulations are also useful. Ground-breaking research in the 1960s by William Condon into the intimate interrelationship between verbal and nonverbal communication confirms this, in that "every sound has a corresponding, instantaneous movement accompanying it, speaker and listener performing a synchronous dance" (Condon, 1980). Studies also reveal that when it comes to showing emotions, people attach particular importance to subtle signs such as gaze or intonation, facial expressions, and body gestures, which all work together to enrich communication. The suggestion is that the Companion's ability to detect these signs in users would considerably improve the interaction, and that people expect this mode of behavior. In this context, the Companion's proactive involvement in the interaction is perceived as a sign of empathy.

Results of the grid, used to measure people's constructs, also confirmed that Samuela was the only Companion considered as a useful artifact, able also to act like a good Companion. She possesses additional features that people considered desirable for a Companion. She was thought of as helpful in everyday life, interactive, emotionally expressive, and good at providing helpful personal advice. Furthermore, people expressed human esthetic preferences as to how they would like their Companion to look and behave.

It is worth noting that some authors have already investigated the effect of anthropomorphic qualities of companions such as race, gender, and sexuality. These features are found to be key to the design of anthropomorphized interfaces. Used as design variables they optimize believability, reliability, and enhance user experience. On the other hand, they can reinforce prejudice and playoff situations of social inequality. For example Mrs. Dewey is an anthropomorphic research engine of Microsoft (Microsoft, 2007) that makes use of the gender and race elements in its interface to ideologically reinforce the Mrs. Dewey website as an area of exotic pleasure and male domination (Sweeney, 2014). These issues should be taken into account when designing anthropomorphized interfaces.

In our study, in users' mind the Companion needs to be female, pretty, a sympathetic listener, have a feminine voice, be chatty, pretty, young, and sexy. As mentioned previously, Samuela is expected to have human behavior as well as agent behavior. Overall, she is expected to be a Companion expressing emotions and showing empathy, which translates to the user as the Companion taking a human-like interest in him.

DEVELOPING AN INTERACTION STRATEGY FOR A VIRTUAL COMPANION

In a further experiment, we invited users to evaluate adjectives to see how they would affectively describe their perfect virtual Companion. Empathy and consistency were found to be the two most important features. People were asked about the limitations of the proposed features in order to best describe their expectations. The importance of reflecting emotions was a significant result of these tests. The majority of test subjects stated that an ECA whose face could not display emotions would feel like having a mere machine in their home. This indicates the impact a Companion's displaying emotions has on its acceptance by users.

Results of this test also show that people expected two levels of behavior from a Companion: emotional and informative-communicative. The first relates to the impact on the user's emotional state, that is, the Companion must be empathetic and display sympathy and cheerfulness. The second focuses on the information that the Companion provides, meaning that they would expect Samuela to be consistent, but not repetitive.

General expectations of a Companion are the display of human qualities such as interaction, humor, contextualization, communication, and involvement, with corresponding physical aspects such as body language, stance, facial expressions, use of space, and gesticulations. Users appreciate companions attempting to exhibit human qualities, fully aware that they do not necessarily have to believe these to be human qualities. Furthermore, for a Companion to be accepted as one, the important feature is that it appears to be empathetic, interact with users by voice, and have some utility.

REFERENCES

Adams-Webber, J. R., & Kelly, G. (1979). *Personal construct theory: Concepts and applications.* New York: Wiley.
Armstrong, K. (1994). *A history of God.* London, England: Random House LLC.
Aylett, R., & Paiva, A. (2012). Computational modelling of culture and affect. *Emotion Review, 4*(3), 253–263.
Banks, M. R., Willoughby, L. M., & Banks, W. A. (2008). Animal-assisted therapy and loneliness in nursing homes: Use of robotic versus living dogs. *Journal of the American Medical Directors Association, 9*(3), 173–177.
Baron-Cohen, S. (2011). *Zero degrees of empathy: A new theory of human cruelty.* UK: Penguin.
Bartal, I. B. A., Decety, J., & Mason, P. (2011). Empathy and pro-social behavior in rats. *Science, 334*(6061), 1427–1430.
Barthes, R. (1965). *Éléments de sémiologie.* Paris: Denoël/Gonthier Editors.
Bates, J. (1994). The role of emotion in believable agents. *Communications of the ACM, 37*(7), 122–125.
Batson, C. D., Duncan, B. D., Ackerman, P., Buckley, T., & Birch, K. (1981). Is empathic emotion a source of altruistic motivation? *Journal of Personality and Social Psychology, 40*(2), 290.
Batson, C. D., Sager, K., Garst, E., Kang, M., Rubchinsky, K., & Dawson, K. (1997). Is empathy-induced helping due to self-other merging? *Journal of Personality and Social Psychology, 73*(3), 495.
Benyon, D., Hansen, P., & Webb, N. (2008). Evaluating human-computer conversation in companions. In *Proceedings of 4th international workshop on human-computer conversation.*
Benyon, D., & Mival, O. (2007). Introducing the companions project: Intelligent, persistent, personalised interfaces to the internet. In *Proceedings of the 21st British HCI group annual conference on people and computers: HCI... But not as we know it—Volume 2* (pp. 193–194). UK: British Computer Society.
Benyon, D., & Mival, O. (2008). Landscaping personification technologies: from interactions to relationships. In *CHI'08 extended abstracts on human factors in computing systems* (pp. 3657–3662). ACM.
Benyon, D., & Mival, O. (2013). Scenarios for companions. In R. Trappl (Ed.), *Your virtual butler* (pp. 79–96). Berlin Heidelberg: Springer.
Bickmore, T. W. (2003). *Relational agents: Effecting change through human-computer relationships.* Massachusetts: Massachusetts Institute of Technology.
Bickmore, T., & Cassell, J. (2000). How about this weather? Social dialogue with embodied conversational agents. In *Proceedings AAAI fall symposium on socially intelligent agents.*
Bickmore, T. W., & Picard, R. W. (2005). Establishing and maintaining long-term human-computer relationships. *ACM Transactions on Computer-Human Interaction (TOCHI), 12*(2), 293–327.
Boukricha, H., Wachsmuth, I., Carminati, M. N., & Knoeferle, P. (2013). A computational model of empathy: Empirical evaluation. In *Proceedings affective computing and intelligent interaction ACII* (pp. 1–6).

Brunt, J. H. (1985). An exploration of the relationship between nurses' empathy and technology. *Nursing Administration Quarterly*, *9*(4), 69–78.
Burleson Daviss, W., Birmaher, B., Melhem, N. A., Axelson, D. A., Michaels, S. M., & Brent, D. A. (2006). Criterion validity of the Mood and Feelings Questionnaire for depressive episodes in clinic and non-clinic subjects. *Journal of Child Psychology and Psychiatry*, *47*(9), 927–934.
Burleson, W., & Picard, R. (2007). Affective learning companions. *Educational Technology-Saddle Brook Then Englewood Cliffs*, *47*(1), 28.
Byrne, R., Lee, P. C., Njiraini, N., Poole, J. H., Sayialel, K., Sayialel, S., et al. (2008). Do elephants show empathy? *Journal of Consciousness Studies*, *15*(10-11), 10–11.
Chou, C. Y., Chan, T. W., & Lin, C. J. (2003). Redefining the learning companion: The past, present, and future of educational agents. *Computers & Education*, *40*(3), 255–269.
Condon, W. S. (1980). The relation of interactional synchrony to cognitive and emotional processes. In M. R. Key (Ed.), *The relationship of verbal and nonverbal communication* (pp. 49–65). The Hague: Mouton Publishers.
Coplan, A., & Goldie, P. (2011). *Empathy: Philosophical and psychological perspectives*. New York: Oxford University Press.
Cowie, R. (2010). Companionship is an emotional business. In Y. Wilks (Ed.), *Close engagements with artificial companions: Key social, psychological, ethical and design issues* (vol. 8, pp. 169–172). Amsterdam: John Benjamins.
Craig, P., Roa-Seïler, N., Díaz, M. M., & Rosano, F. L. (2013). The role of embodied conversational agents in collaborative face to face computer supported learning games. In *Proceedings of the 26th international conference on system research, informatics & cybernetics, Baden Baden, Germany*. Winner of the Outstanding Scholarly Contribution Award.
Crossley, L. (2003). Building emotions in design. *The Design Journal*, *6*(3), 35–45.
Davis, M. H. (1980). A multidimensional approach to individual differences in empathy. *Psychology*, *10*, 85.
Davis, M. H. (1994). *Empathy: A social psychological approach*. Boulder, US: Westview Press.
Decety, J., & Jackson, P. L. (2006). A social-neuroscience perspective on empathy. *Current Directions in Psychological Science*, *15*(2), 54–58.
Decety, J., & Lamm, C. (2006). Human empathy through the lens of social neuroscience. *The Scientific World Journal*, *6*, 1146–1163.
Deegan, P. (1999). Kill Clippy. ZDNet.com. Retrieved November 22, 1999 from http://www.zdnet.com/products/stories/reviews/0,4161,2226725-1,00.html.
Edgell, S. (1972). Marriage and the concept companionship. *British Journal of Sociology*, *23*, 452–461.
Eisenberg, N. (2000). Emotion, regulation, and moral development. *Annual Review of Psychology*, *51*(1), 665–697.
Ellison, M., & Spiegel, A. (2013). Her [Film]. United States: Warner Bros. Pictures.
Fomichov, V. A., & Fomichova, O. S. (2014). An Imperative of a poorly recognized existential risk: Early socialization of smart young generation in information society. In V. Klyuev (Ed.), *Special issue: Advances in semantic information retrieval* (vol. 38, pp. 59–70).
Fransella, F., Bell, R., & Bannister, D. (2004). *A manual for repertory grid technique* (2nd ed.). England: Academic Press.
Gleitman, L., & Papafragou, A. (2005). Language and thought. In K. Holyoak & R. Morrison (Eds.), *The cambridge handbook of thinking and reasoning* (p. 858). New York: Cambridge University Press.
Goldman, A. I. (1993). Ethics and cognitive science. *Ethics: An International Journal of Social, Political, and Legal Philosophy*, *103*, 337–360.
Gorry, G. A., & Westbrook, R. A. (2011). Once more, with feeling: Empathy and technology in customer care. *Business Horizons*, *54*(2), 125–134.

Greimas, A. J. (1968). Conditions d'une sémiotique du monde naturel. *Langages, 10,* 3–35. Armand Colin.
Heirman, W., & Walrave, M. (2008). Assessing concerns and issues about the mediation of technology in cyberbullying. *Cyberpsychology: Journal of Psychosocial Research on Cyberspace, 2*(2), 1–12.
Hinduja, S., & Patchin, J. W. (2010). Bullying, cyberbullying, and suicide. *Archives of Suicide Research, 14*(3), 206–221. Taylor & Francis Group.
Hjort af Ornäs, V. (2010). *The significance of things: Affective user-artefact relations.* Gothenburg, Sweden: Chalmers University of Technology.
Hoffman, M. L. (1982). Development of prosocial motivation: Empathy and guilt. *The Development of Prosocial Behavior, 281,* 313.
Holland, N. N. (2003). *The willing suspension of disbelief: A neuro-psychoanalytic view.* PSYART. Retrieved from, http://www.psyartjournal.com/article/show/n_hollandthe_willing_suspension_of_disbelief_a_ne.January.
Hollingsworth, L. A., Didelot, M. J., & Smith, J. O. (2003). Reach beyond tolerance: A framework for teaching children empathy and responsibility. *The Journal of Humanistic Counseling, Education and Development, 42*(2), 139–151.
Hurd, G. A., (Producer), & Cameron, J. (Director), (1984). *The Terminator [Film].* United States: Metro Goldwyn Mayer.
Ickes, W. J. (1997). *Empathic accuracy.* New York: The Guilford Press.
Joplin, J. R. W., & Daus, C. S. (1997). Challenges of leading a diverse workforce. *The Academy of Management Executive, 11*(3), 32–47.
Kelly, G. (2003). *The psychology of personal constructs: Volume two: Clinical diagnosis and psychotherapy.* UK: Routledge.
Kubrick, S., (Producer), & Kubrick, S. (Director), (1968). *2001: A Space Odyssey [Film].* United Kingdom: Metro-Goldwyn-Mayer.
Lawson, B. (2007). *Language of space.* London, England: Routledge.
LeDoux, J. E. (1995). Emotion: Clues from the brain. *Annual Review of Psychology, 46*(1), 209–235.
Luehning, E. (2001). *Microsoft tool Clippy gets a pink slip.* CNET News. Retrieved from, http://news.cnet.com/2100-1001-255671.html.
Luh, D. B., Li, E. C., & Kao, Y. J. (2015). The development of a companionship scale for artificial pets. *Interacting with Computers, 27*(2), 189–201.
Maller, P., Salamon, A., Mira, N., Boggio, A., Giro, J., Cuozzo, J., et al. (2012). New practices to structure and elicit improvement opportunities in scientific software development teams. In *12th international conference on computational science and its applications* (pp. 121–124).
Marsella, S., & Gratch, J. (2014). Computationally modeling human emotion. *Communications of the ACM, 57*(12), 56–67.
Maslow, A. H. (1943). A theory of human motivation. *Psychological Review, 50*(4), 370.
McNicholas, J., & Collis, G. M. (2001). Children's representations of pets in their social networks. *Child: Care, Health and Development, 27*(3), 279–294.
McQuiggan, S. W., & Lester, J. C. (2007). Modeling and evaluating empathy in embodied companion agents. *International Journal of Human-Computer Studies, 65*(4), 348–360.
Microsoft (2007). Ms. Dewey: Best of Dewey. Podcast retrieved from http://www.criticalcommons.org/Members/ccManager/clips/ms-dewey-best-of-dewey/view.
Miller, P. A., & Eisenberg, N. (1988). The relation of empathy to aggressive and externalizing/antisocial behavior. *Psychological Bulletin, 103*(3), 324.
Mori, M. (1970). The uncanny valley. *Energy, 7*(4), 33–35.
Muller, M. J., Wharton, C., McIver, W. J., Jr., & Laux, L. (1997). Toward an HCI research and practice agenda based on human needs and social responsibility. In *Proceedings of the ACM SIGCHI conference on human factors in computing systems* (pp. 155–161). New York: ACM.

Neimeyer, R. A., & Bridges, S. K. (2004). *The internet encyclopedia of personal construct psychology*. Retrieved from, http://www.pcp-net.org/encyclopaedia/pc-theory.html.
Nishida, T. (2013). Toward mutual dependency between empathy and technology. *AI & Society, 28*(3), 277–287.
Nissenbaum, H., & Walker, D. (1998). Will computers dehumanize education? A grounded approach to values at risk. *Technology in Society, 20*(3), 237–273.
Norman, D. A. (2002). Emotion and design: Attractive things work better. *Interactions Magazine, 11*, 36–42.
Osgood, C. E., Suci, G. S., & Tannenbaum, P. H. (1957). *In the measurement of meaning*. Urbana, IL: University of Illinois.
Paiva, A., Dias, J., Sobral, D., Aylett, R., Woods, S., Hall, L., et al. (2005). Learning by feeling: Evoking empathy with synthetic characters. *Applied Artificial Intelligence, 19*(3-4), 235–266.
Picard, R. W., & Klein, J. (2002). Computers that recognize and respond to user emotion: Theoretical and practical implications. *Interacting with Computers, 14*(2), 141–169.
Pierce, J. (2008). Mice in the Sink. *Environmental Philosophy, 5*(1), 75–96.
Plutchik, R. (2001). Integration, differentiation, and derivatives of emotion. *Evolution and Cognition, 7*(2), 114–125.
Prendinger, H., & Ishizuka, M. (2005). The empathic companion: A character-based interface that addresses users' affective states. *Applied Artificial Intelligence, 19*(3-4), 267–285.
Pulman, S. G., Boye, J., Cavazza, M., Smith, C., & De La Cámara, R. S. (2010). How was your day? In *Proceedings of the 2010 workshop on companionable dialogue systems* (pp. 37–42). Uppsala, Sweden: Association for Computational Linguistics.
Reeves, B., & Nass, C. (1996). *How people treat computers, television, and new media like real people and places*. Stanford, CA: CSLI Publications and Cambridge University Press.
Reif, J. H., & Wang, H. (1999). Social potential fields: A distributed behavioral control for autonomous robots. *Robotics and Autonomous Systems, 27*(3), 171–194.
Roa Seïler, N., & Benyon, D. (2010). Designing companions with Kansei. In *Proceedings of international conference on Kansei engineering and emotion research, Paris, France*.
Roa-Seïler, N., Benyon, D., & Leplâtre, G. (2009). An affective channel for companions. In *Electronic proceedings empathic agents workshop at AAMAS, Proc. of 8th Int. Conf. on autonomous agents and multiagent systems, Budapest, Hungary*.
Roa-Seïler, N., Craig, P., Aguilar, J., Saucedo, A. B., Díaz, M. M., & Rosano, F. L. (2014). Defining a child's conceptualization of a virtual learning companion. In *Proceedings of international technology, education and development conference INTED, Valencia, Spain*, ISBN: 978-84-616-8412-0.
Rodrigues, S. H., Mascarenhas, S., Dias, J., & Paiva, A. (2015). A process model of empathy for virtual agents. *Interacting with Computers, 27*(4), 371–391.
Rohlin, K. (2002). *Apparent usability attributes*. Norwegian University of Science and Technology. Retrieved from, http://www.ivt.ntnu.no/ipd/docs/pd9_2002/Rohlin_II.pdf.
Rueckert, L., & Naybar, N. (2008). Gender differences in empathy: The role of the right hemisphere. *Brain and Cognition, 67*(2), 162–167.
Russell, J. A. (1980). A circumplex model of affect. *Journal of Personality and Social Psychology, 39*(6), 1161.
Shneiderman, S. B., & Plaisant, C. (2005). *Designing the user interface* (4th ed.). USA: Pearson Addison Wesley.
Sillars, A. L., Weisberg, J., Burggraf, C. S., & Zietlow, P. H. (1990). Communication and understanding revisited married couples' understanding and recall of conversations. *Communication Research, 17*, 500–522.
Silver, J., (Producer), Wachowski, A., & Wachowski, L. (Directors), (1999). *The Matrix* [Film]. United States: Warner Bros.

Simpson, J. A., & Weiner, E. S. C. (1989). *The Oxford English dictionary: Vol. 2.* Oxford, New York: Clarendon Press.

Slonje, R., & Smith, P. K. (2008). Cyberbullying: Another main type of bullying? *Scandinavian Journal of Psychology, 49*(2), 147–154.

Smith, C., Crook, N., Boye, J., Charlton, D., Dobnik, S., Pizzi, D., et al. (2011). Interaction strategies for an affective conversational agent. *Presence: Teleoperators and Virtual Environments, 20*(5), 395–411.

Smith, P. K., Mahdavi, J., Carvalho, M., Fisher, S., Russell, S., & Tippett, N. (2008). Cyberbullying: Its nature and impact in secondary school pupils. *Journal of Child Psychology and Psychiatry, 49*(4), 376–385.

Sweeney, M. (2014). *Not just a pretty (inter) face: A critical analysis of Microsoft's' Ms. Dewey.* Doctoral dissertation, University of Illinois at Urbana-Champaign.

Tan, F. B., & Tung, L. (2003). Exploring website evaluation criteria using the repertory grid technique: A web designers' perspective. In *Proceedings of SIGHCI.*

Vihma, S. (2003). On actual semantic and aesthetic interaction with design objects. In *Proceedings of 5th European academy of design conference: Techne design* (pp. 1–7).

Vugt, H. C. V., Bailenson, J. N., Hoorn, J. F., & Konijn, E. A. (2010). Effects of facial similarity on user responses to embodied agents. *ACM Transactions on Computer-Human Interaction (TOCHI), 17*(2), 7.

Wattles, J. (1997). *The golden rule.* New York: Oxford University Press.

Williams, K. J., Peters, J. C., & Breazeal, C. L. (2013). Towards leveraging the driver's mobile device for an intelligent, sociable in-car robotic assistant. In *Intelligent vehicles symposium (IV)* (pp. 369–376).

Woolf, B., Burleson, W., Arroyo, I., Dragon, T., Cooper, D., & Picard, R. (2009). Affect-aware tutors: recognizing and responding to student affect. *International Journal of Learning Technology, 4*(3), 129–164.

CHAPTER 5

Spoken Dialog Agent Applications using Emotional Expressions

Kaoru Sumi
Future University Hakodate, Hakodate, Japan

INTRODUCTION

An increasing number of people are in need of mental care, because associated social issues include problems of bullying in schools, mental disability and hikikomori (a Japanese term to refer to the phenomenon of reclusive adolescents or young adults who withdraw from social life), or social withdrawal of young people, suicide of young people, people being pressured to quit work in middle age, and PTSD (posttraumatic stress disorder) of the victim.

This chapter introduces a study on a spoken dialog agent system for mental care. There are many studies of therapeutic systems that have nonverbal interfaces as represented by robot therapy. These include Paro (Wada & Shibata, 2007), which is a seal-type robot developed by AIST in Japan, a bear-type social robot developed by Fijitsu Research Laboratory (http://www.youtube.com/watch?v=AwWeN1ARy74), dog-type robots of Pooch (http://www.robotsandcomputers.com/robots/manuals/Poo-Chi.pdf) developed by Sega, and AIBO (http://www.sony.jp/products/Consumer/aibo/) developed by Sony. These are used as therapeutic toys and have nonverbal interfaces.

I think that a system which offers intelligent conversation not only heals but can make indepth contact with a person. In addition, among people who are hikikomori or social withdrawal at home tend to behave as staying in front of computer most of the time. I think information and communication technologies (ICTs) can lead to the potential for support for hikikomori or social withdrawal among people. However, studies of systems that feature conversation to support mental health are less well understood in Japan.

On the other hand, in the United States and Europe, mental health care research using ICTs has officially begun. Defense Advanced Research

Projects Agency (DARPA) supports the PTSD treatment of returned soldiers using remote systems, and promotes research and development of Detection and Analysis of Psychological Signals (DCAPS) of next-generation ICT, and artificial intelligence. In the developed European nations, great investment has been made into research and development of psychosocial support using ICTs.

Our research group has studied persuasive technology via human agent interaction using facial expressions and dialogs, with the goal of developing a virtual agent that can persuade a human to act in a certain way (Sumi, 2012; Sumi & Nagata, 2010, 2012).

According to Media Equation (Reeves & Nass, 1996), people tend to respond to media as they would to another person. According to persuasive technology (Fogg, 2003), if a user recognizes the presence of something in a computer, he or she will respond to it according to the normal social rules. However, there are still many things that we do not know about how an agent's response affects a user during their interaction. I think systems can change users' behavior and this kind of research is related to the area of counseling by systems.

In the past, a computer program and an early example of primitive natural language processing, ELIZA (Weizenbaum, 1966) was developed to imitate a Rogerian psychotherapist, and was an early example of primitive natural language processing. SimCoach (Rizzo et al., 2011) is a spoken dialog agent system used in the mental care of returned soldiers. It displays a speaking counselor agent and analyzes the psychology of the user using a question and answer technique. Docpal (http://www.hellodocpal.com/) from Orange Silicon Valley is a remote medical care system using augmented reality, KINECT, speech recognition, text to speech, avatar interaction, and a decision support system. Remote medical care is very important as mental care because there are many people who do not readily go to the hospital. We aim toward the development of a mental care system that can be used easily in house.

This chapter introduces a spoken agent system for mental care that uses expressive facial expressions and positive psychology (Peterson, 2006; Seligman, 2004). In the following, in section "Experiment Investigating Impressions and Behavior Change Caused by Replies from the Agent," the chapter describes an experiment investigating the effect that the expression and words of the agent have on people, compares the effect that the expression and words of the agent and robot have on people in section "Comparative Experiment on Effects of a Virtual Agent and a Robot," introduces a spoken agent for customer services using expressive facial expressions in section "A Spoken Agent System for Learning Customer Services," introduces a spoken agent for mental care using expressive facial expressions and positive psychology in section "A Spoken Agent System for Mental Care using Expressive Facial Expressions and Positive Psychology," then presents a "Discussion" and ends with "Conclusion."

EXPERIMENT INVESTIGATING IMPRESSIONS AND BEHAVIOR CHANGE CAUSED BY REPLIES FROM THE AGENT

In this section, we introduce an experiment investigating the impressions and persuasive effect of the replies from the agent, and consider the facial and word expression used in the agent's replies. We conducted the experiment on how the user felt about the agent's reaction by setting up an emotion-arousing scenario for the user (Sumi & Nagata, 2010). A total of 1236 men and women participated in the research. The synchronization of the agent's words with the user's emotions has a major impact on the impression of the agent as perceived by the user. However, the synchronization of facial expressions of the agent with the user's emotion does not have a major impact on the creation of an impression.

Experimental Method

We chose six kinds of feelings. From the total of 216 combinations, covering multiple feelings that the user felt (6 patterns) and the facial expressions for the agent's interaction with the user (6 patterns) and word expressions used by the agent (6 patterns), we selected 96 patterns in this experiment. These covered 16 patterns in each feeling: empathetic words and consistent facial expressions, nonempathetic words and consistent facial expressions, word consistent and facial inconsistent, and word inconsistent and facial consistent. This is because the conditions of nonempathetic and both inconsistent word and facial expressions are nonsensical in normal communication. This is the condition where the word and facial expressions are inconsistent, which is the condition for double bind communication, but it can be considered as either word or facial being empathetic to the user. The case of nonempathetic condition and inconsistent word and facial expressions can be considered as pathological.

A total of 1236 people, 568 male and 668 female (AV. age 38.0, SD age 11.5), were assigned 96 contents. More than 10 users were assigned to each content.

Experimental materials are as follows.
- Six kinds of emotion-arousing scenarios
  "Joy," "anger," "sadness," "disgust," "fright," and "surprise" scenarios were selected as scenarios with a high concordance rate in a preliminary experiment as emotion-arousing scenarios for each emotion. These scenarios were described by a male reader reading in a neutral manner.
- Agent
  - Facial expressions
    Faces representing "joy," "anger," "sadness," "disgust," "fright," and "surprise" were selected as faces with a high concordance rate in a preliminary experiment as emotional faces.
  - Word expressions
    The agent dialog was read by a female reader with emotions conveyed. At first, as empathetic dialog, "I think so, too" or "I don't think so" as nonempathetic dialog were spoken. Then, emotionally, the dialog of "that's nice," "that's aggravating," "that's sad," "that's disgusting," "that's scary," and "that's a big surprise" were spoken.
- Impression evaluation task
  In this experiment, we use nine factors: three factors for interpersonal impression evaluation (Daibo, 1978) were
  - "affable-inaffable,"
  - "serious-unserious,"
  - "conversable-inconversable"
  and original six factors
  - "reliable-unreliable,"
  - "gentle-bitter,"
  - "egotistic-humble,"
  - "empathetic-unempathetic,"
  - "authoritative-unauthoritative," and "offensive-inoffensive."

We prepared 96 contents to cover the combination of emotions that a user feels, the facial expressions of the agent, and the word expressions used by the agent. These contents were developed using the Bot3D Engine (http://www.atom.co.jp/bot/), which displays an agent on web pages. The Bot3D Engine is an embedded engine for developing software using the Web3D plug-in and ActiveX component. Users can use the 3D agent program only to access web pages that have the program embedded.

The examination was conducted in the form of a questionnaire on the web. The content was displayed on user's own PC monitor after the user accessed the target URL.

1. Inputting the user's attributes
   The users were asked about their sex, age, marital status, occupation, intended purpose of using the PC, and for how long they had been using a PC.
2. Instruction
   Next, the following teaching sentences were presented.

   This examination aims to discover what emotions people feel in various cases. There is no correct answer, so please say exactly what you think and feel. This examination is not a test of your personal abilities. The answers will be analyzed statistically and private information will not be released. First, please consider the given scenario and then select from the alternatives the emotions that you feel. Next, an animated character will respond to your selected answer. Please answer the question by giving your impression of the character. Your answer should relate only to this scenario. Please do not include feeling from previous scenarios, but think scenario by scenario.

3. Content presentation
   Each user was presented with 1 of 96 contents. One of the emotion-arousing scenarios was read by a male voice. It then asked: "What kind of emotion do you feel?" and prompted the user to select from the alternatives "joy," "anger," "sadness," "disgust," "fright," and "surprise." On the other hand, the female agent on the screen responded with facial and word expressions.
4. Evaluation of the impression given by the agent
   Users were asked: "How do you feel about this person? Please answer using the degrees listed in the questionnaire." Five conditional moods in nine answers, "conversable-inconversable," "reliable-unreliable," "gentle-bitter," "egotistic-humble," "empathetic-unempathetic," "authoritative-unauthoritative," "offensive-inoffensive," "serious-unserious," and "affable-inaffable" were given and the user selected a suitable answer. The order of the terms was kept constant throughout the questions.

Results and Discussion

People are easily persuaded when the agent was favorable impression. This result is very natural; however, the combination of favorable was different from our prediction. First, we predicted that words and facial expressions reflected on the emotions aroused by the scenario would lead to the most favorable impression, so we set these data as the control group. In fact, there were more favorable impressions than those obtained for the control group. For example, the words and facial expressions were "joy" when the user's emotion was "joy" for the control group. It is very interesting that when the user's emotion was "joy," the agent's words for "joy" with facial expressions of "surprise," "sadness," or "fright" were most favorable. On the other hand, when the user's emotion was "fright," the agent's words for "fright" with facial expressions of "disgust" or "sadness" were the most favorable.

These facial expressions were recognized as the emotion conveyed by the words and were more empathetic and somewhat meaningful emotions. For example, when the user's emotion was "joy," the agent's words of "joy" with facial expressions of "surprise" or "fright" might have been recognized as the agent being exaggeratedly surprised at the "joy" scenario. When the user's emotion was "joy," the agent's words of "joy" with facial expressions of "sadness" might have been recognized as the agent being highly pleased from the heart at the "joy" scenario. When the user's emotion was "fright," the agent's words of "fright" with facial expressions of "sadness" might have been recognized as the agent grieving deeply at the user's "fright" scenario. When the user's emotion was "fright," the agent's words of "fright" with facial expressions of "disgust" might have been recognized as the agent feeling deep hate at the user's "fright" scenario.

Through these observations, we concluded that there is a rule for facial expressions: in a certain scenario, synchronizing foreseen emotion of the user caused by the situation will make a favorable impression. For example, when the user has the emotion of "joy," he/she wants someone to be surprised or highly pleased. Then, showing a surprised or highly pleased face expression makes the user feel favorable impression. When the user has the emotion of "fright," he/she wants someone to grieve deeply or disgust. Then, showing grieved or disgust face expression make the user feel a favorable impression. Users want the agent to synchronize their foreseen emotion by hearing the news instead of simply showing synchronized reaction according to emotion at present time.

COMPARATIVE EXPERIMENT ON EFFECTS OF A VIRTUAL AGENT AND A ROBOT

In this section, we introduce a comparative experiment investigating the effect of a virtual agent and a robot. We describe the possible future use of agents and robots as persuasive interaction partners. We conducted a comparative experiment to examine what kind of talker is more appropriate for persuasive interaction, namely, human–agent interaction or human-robot interaction (Sumi & Nagata, 2012). In this case, we set up a situation of persuasive interaction to sustain motivation.

Experimental Method

In order to perform this comparison between types of interaction, we chose three factors, namely character condition (virtual agent/robot), emotional condition (emotional/emotionless) in terms of facial & word expressions, and encouraging condition (encouragement/without encouragement) using words during the user's work. A total of 170 people, 44 male and 126 female graduate and undergraduate students, were assigned eight kinds of conditions. We used a female virtual agent for interacting with the user. In accordance with the previous experiment, we chose favorable facial and word expressions. Joyful and emotionless faces were used for the agent. At first, as empathetic dialog, "I think so, too" was spoken or "I don't think so" as nonempathetic dialog. Then, in the case of the emotional face, the dialog of "that's nice" was spoken emotionally. In the case of the emotionless face, the dialog of "that's nice" was spoken emotionlessly.

The animation contents were created using java script and .wmv files. In the case of virtual agent condition, the content is displayed by note PC in front of the subject. The content was projected onto a robot's face using a cinematographic projector in the case of the robot condition. The face of the robot, Robovie-R3, was covered with a screen cloth. Subjects could not see the projector because a partition shielded it from them. A virtual agent and a robot had the same facial expression.

During the user's trial of the task, a talker encouraged the user with the phrases of "Ganbatte!" ("Do your best!") and "Yoi choshi desune!" ("Good job!").

Procedure is as follows.
1. Instruction and inputting the user's attributes
    The users were asked about their sex and age. The following teaching sentences were presented.

    This examination aims to study the attitude that people feel about communication. There is no correct answer, so please say exactly what you think and feel. This examination is not a test of your personal abilities. The answers will be analyzed statistically and private information will not be released.

    First, please look at the computer display (or the robot). This character will explain several things to you. An experimenter will sometimes make a supplementary statement along the way. In that case, please follow his statement. Let's start.

2. Emotion arousal and controlling emotional valence
    "Please consider the given scenario and then select from the alternatives the emotions that you feel. Next, an animated character will respond to your selected answer." Then, the scenario was read.
3. Content presentation
    Each user was assigned to 1 of 16 conditions. Joyful scenarios were read by a male voice. The voice then asked: "What kind of emotion do you feel?" and prompted the user to select from the alternatives "joy," "anger," "sadness," "disgust," "fright," and "surprise." Then, the character responded with facial and word expressions.
4. Introduction to a task of filling in pictorial figures
    Then, they worked at a simple task to determine whether they kept motivation while doing it through the presence or absence of the agent's favorable impression or encouragement. The following teaching sentences were presented by a male voice. "Latest research reveals that it is effective by working at a task of filling pictorial figures every day to keep exquisite our thinking power for all time. A task of filling pictorial figures means the task for filling small pictorial figures in a predetermined order as fast as you can in a careful manner. To achieve an effect, the higher the number of pictorial figures to fill, the better. Please attempt this task immediately." The character asks, "Do you want to work at the task of filling pictorial figures?" as Question 1.
5. Conducting a task of filling in pictorial figures
    If the subject answered yes to Question 1, then the character said, "Oh, you want to do it. Let's try to improve your thinking power."

    If the subject answered no to Question 1, then the character said, "Oh, you don't want to do it. But, let's try to improve your thinking power." Then, they started the task. At that time, emotional valence of the character was controlled according to the conditions.

    The character said, "This is a form for filling in pictorial figures. When you have finished filling in 10 figures, click the button for "finished 10 figures" or call me.

    So, let's start." The subject was told to fill in several printed pictorial figures (Figure 5.2) using designated colors. In the case of the encouraging condition, the character repeated "Ganbatte!" ("Do your best!") or "Yoi choshi desune!" ("Good job!") occasionally during the task. In the case of without-encouraging condition, the character kept silent during the task.

When the subject finished filling in 10 figures, the character said "Otsukare sama deshita" ("Well done"). "You finished ten figures now. Will you keep filling in the next ten figures?" using emotionless facial and word expressions as Question 2, as requiring a choice of carrying on the task. If the subject answered yes, he carried on the task of filling in 10 more figures. If the subject answered no, the character requested "Oh, you don't want to do it. However, the more pictorial figures that you fill in, the better it is for your thinking power. Would you like to keep working on it?" If the subject still answered no, then the character enlightened "Ostukare sama deshita." (Well done.), "Let's finish on this" and then finished the task. At that time, emotional valence of the character was controlled according to the conditions.

When the subject finished filling all of the 20 trials of 10 figures, the character enlightened "Ostukare sama deshita." ("Well done.") "Let's finish on this," then finished the task.

6. Questions about the task and the character

Next, the subject was asked questions about the task and the character. The questions were as follows.

- Question 3: Did you enjoy filling in the pictorial figures?
- Question 4: Do you think that you raised your thinking power?
- Question 5: Do you think that you want to continue this task after tomorrow?
- Question 6: How did you accept the character's explanation of effect on filling pictorial figures?
- Question 7: Did the character's encouragement motivate you during the task?
- Question 8: Character's impression

Users were asked: "How do you feel about this person? Please answer using the degrees listed in the questionnaire." Five conditional moods in nine answers, "conversable-inconversable," "reliable-unreliable," "gentle-bitter," "egotistic-humble," "empathetic-unempathetic," "authoritative-unauthoritative," "offensive-inoffensive," "serious-unserious," and "affable-inaffable" were given and the user selected a suitable answer.

7. Debriefing

Finally, the character disclosed to the user that the true aim of this examination was to clarify the user's impression of the character. The information about the effect on filling in pictorial figures was just a diversion.

Results and Discussion

As a result of our investigation into what kind of talker is preferable for persuasion, we obtained the following insights. Characteristics of a virtual agent and a robot were as follows. Virtual agent: The number of filling figures was greater than the robot condition. Subjects who thought that their thinking power had been increased were greater in number than those with the robot condition. That is to say, the virtual agent is more persuasive than the robot. Robot: A robot was experienced as being gentle by users. A robot was especially taken as being gentle when it had emotional expression and encouraged a user. A robot was taken as inoffensive. A robot was especially taken as being inoffensive when it had emotional expression. A robot was humble when it encouraged a user.

Overall, a virtual agent was more influential on the user than a robot in situations where maintenance of motivation was needed. In particular, in the case of emotional and encouragement condition, when the user who once answered "no" tended to change their mind to keep working on the task, if the user persuaded by the agent tended to change their mind, however the user persuaded by the robot tended not to change their mind. This is a very interesting finding.

Summary of comparison between emotional and unemotional is as follows. Emotional: Emotional was taken more plausible such as believing talk of improving user's thinking power than emotionless. Emotional was gentle, empathetic, humble, and affable. When the character is a robot, emotional is humble. Emotionless: Emotional and without encouragement makes authority.

Summary of comparison between encouragement and without encouragement is as follows. Encouragement: In the case of robot, encouragement was enjoyable. In the case of virtual agent, emotionless and encouragement was enjoyable. Encouragement was gentle and humble. Without Encouragement: Without encouragement motivated the users. Especially, without encouragement and emotional motivated the users. Without encouragement made authoritative.

A virtual agent and a robot in these experiments expressed only facial expressions and word expressions; however, if they would be able to make other expressions, such as gestures or other movements, the result might be different. Additionally, since these experiments were conducted only in Japan and subjects were all Japanese residents, further examination is required, taking into account cultural differences.

A SPOKEN AGENT SYSTEM FOR LEARNING CUSTOMER SERVICES

From the result of the first experiment, it is thought that an agent can make in-depth contact with the user by foreseeing the user's emotion and empathizing with the user. From the result of the next experiment, it is thought that the agent should have rich emotional expression and should not encourage the user more than necessary. According to the results, we developed the agent system for learning facial expressions and words of customer services.

This first character agent system consists of a 3D character agent display system, a speech recognition system, a speech synthesis system, a dialog control system, and a facial expression recognition system. This system is for providing educational training in hospitality through dialog with a character agent. This system focuses on the Japanese style of service-mindedness, which is typified by paying attention to individual customers (Sumi & Ebata, 2013a, 2013b).

The 3D character agent display system displays a 3D character agent, which can show the facial expressions of "smile," "laugh," "anger," "sadness," "disgust," "fright," and "surprise" (Figure 5.1). The system superimposes the mouth shapes of the vowels onto each facial expression,

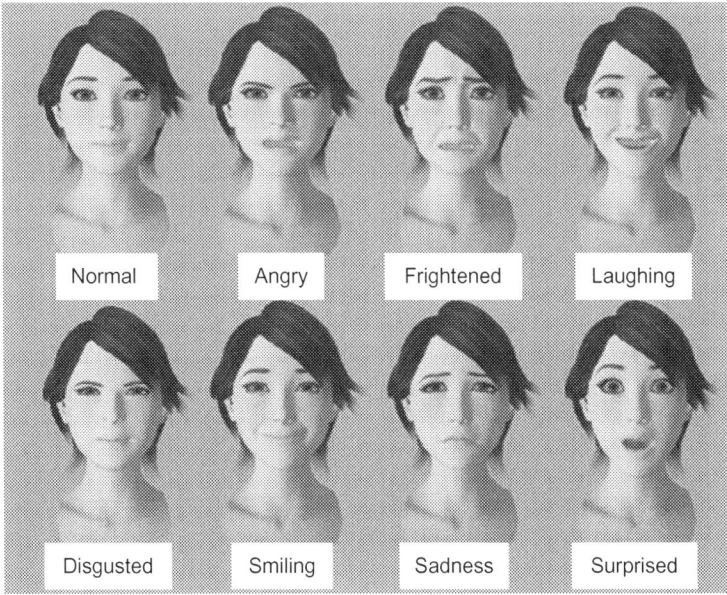

Figure 5.1 Facial expressions.

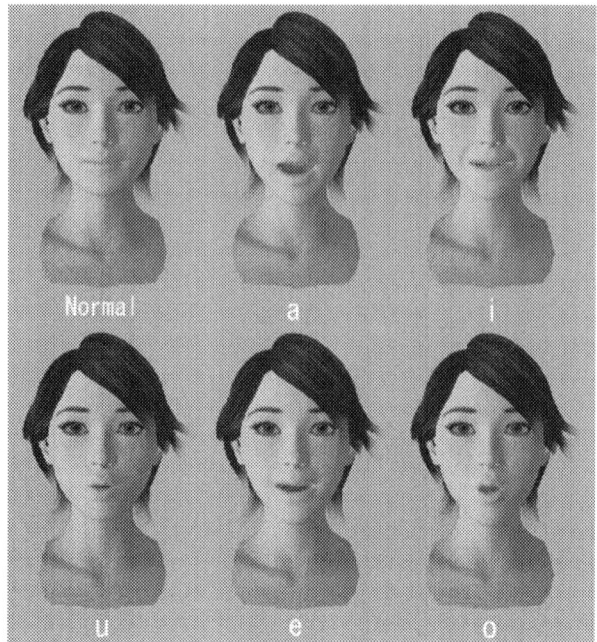

Figure 5.2 The mouth shapes of the vowels.

lip-synching with the sounds (Figure 5.2). Fluid movement of the facial expressions and lip-synching is made possible using Microsoft XNA morphing technology. Our system uses Google speech Application Programming Interface (API) as the speech recognition system, and AITalk as the Japanese speech synthesis system. As a dialog control system, we revised Artificial Intelligence Markup Language (AIML, http://www.alicebot.org/documentation/) which is based on Extensible Markup Language (XML) to use the Japanese language. Using templates, we can express a dialog freely.

<pattern>* tell me your name *</pattern>
<template>My name is Ayaka</template>

For example, if there is a question "Tell me your name?" from a user, then the system answers, "My name is Ayaka," in this case. The important pattern here is "*tell me your name*," so the system can answer the sentence "Please tell me your name" or "Could you tell me your name?" because these sentences include the phrase "tell me your name." When we analyze Japanese language, we have to add spaces between words because there are no spaces in Japanese sentences. We use MeCab, which is a fast and customizable Japanese morphological analyzer, to add spaces between words. As the

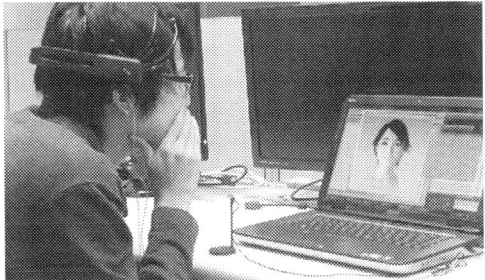 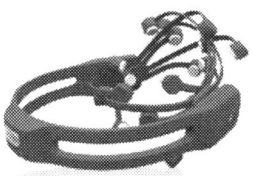

Figure 5.3 Emotiv EPOC and the system.

facial expression recognition system, this system uses the facial recognition application of brain wave measuring equipment called "Emotive EPOC" (Figure 5.3). Emotive EPOC is a brain-computer interface which has 16 electrodes based on electroencephalography (EEG) technology. It recognizes the intensity of facial expressions digitally.

Using our system, a user can talk to the character agent, and the character agent teaches the user how to interact with customers. The character agent is displayed on the screen and sometimes the system displays lines in a scene which the user should practice, along with an appropriate facial expression (Figure 5.4). The system judges whether the user has spoken the lines appropriately by comparing them with the speech recognition system using AIML templates. The system judges whether the user's facial expression is appropriate or not by comparing it with the facial expression recognition system.

Experiment and Results

We conducted an experiment using the developed system on 20 subjects, comprising males and females aged from 21 to 25 years old (AV. age 22.15, SD age 0.73). Subjects operated the system and answered a questionnaire. For the question "Did you enjoy the system?" There were many very favorable comments. The subjects were divided over the answer to the question "Do you have an aversion to attaching a headset?" For example, there were positive answers such as "I feel it was like a hat," and "I feel really comfortable with the headset." On the other hand, there were negative answers such as "I had a pain behind the ear," and "It was a bit hard to attach it." It is thought that the users had a hard time adjusting the headset. For the question "Did the system speak easily?" there were many comments that were critical. For example, "It was hard for my words to be recognized." Improvement of the speech recognition system is needed. There was a comment that "Because the face of the agent was in front of me, I was a little tense."

Figure 5.4 A spoken agent system (The first version).

On the other hand, there were the comments that "I have learned the service by being tense" and "It was more comfortable than talking with a person in a meeting." In this way, we might say that the system utilized the advantage of being a machine. For the question "Was the voice from the system easy to understand?" there were many very favorable comments such as "It is easy to hear it," "The sound is very easy to understand."

A SPOKEN AGENT SYSTEM FOR MENTAL CARE USING EXPRESSIVE FACIAL EXPRESSIONS AND POSITIVE PSYCHOLOGY

We developed a spoken agent system for mental care of hikikomori persons improving the previous system. According to the result of the experiment, it was revealed we must improve interface of the system, especially for the attachment of Emotiv. We improved a spoken agent system. Figure 5.5 is the system framework. The system consists of a 3D character agent display system, a dialog control system, Japanese word spacing system, a speech recognition system, a speech synthesis system, and a facial/gesture expression

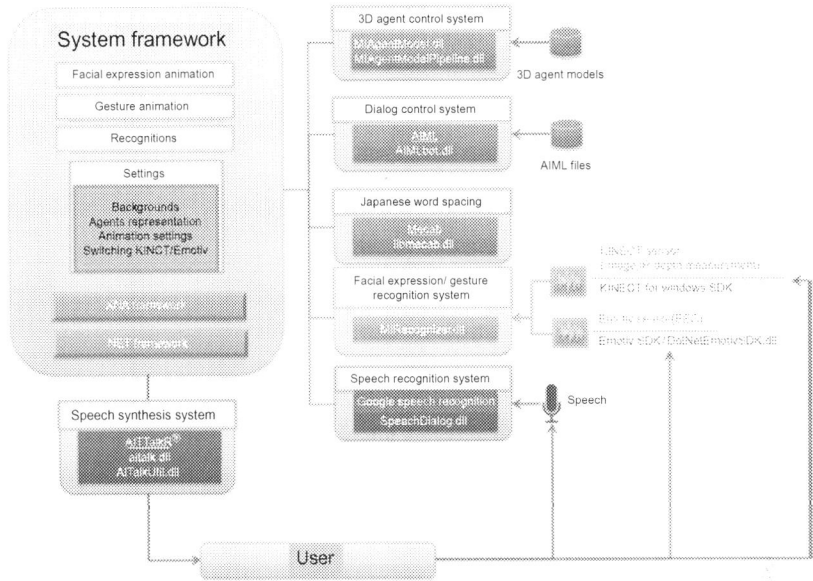

Figure 5.5 System framework.

recognition system. The improved points are as follows. (1) KINECT sensor recognizes a user's facial expressions and gestures; (2) an agent displays gestures (Figure 5.6); (3) the facial expression/gestures can be controlled in intensity digitally (Figure 5.7); (4) an agent can be controlled in its size, direction, and position (Figure 5.8); (5) a background of the system is changeable (Figure 5.9); and (6) an agent blinks its eyes.

Interactive talks using gestures are enabled by recognizing a user's motion from the skeleton using a KINECT sensor. In addition, a simple facial expression is enabled by recognition of a user's face using KINECT sensor without an Emotiv. The system can be switched between a KINECT and/or an Emotiv. An agent displays gestures of "nodding," "nodding no," "waving her hand," "bowing her head," and "bowing her head deeply." An agent can control its size (full body/upper body/face or parentage of full body), direction and position using the dialog editor of figure. The background of the system can be changed using the dialog editor. The agent blinks its eyes for making the agent more human. Because, according to the experiment, there are many comments that the agent should blink its eyes.

In addition, this system offers the function of answering positively when a user speaks negatively, for creating a positive mood. This system answers with positive words to a user's negative words using Negapo Dictionary

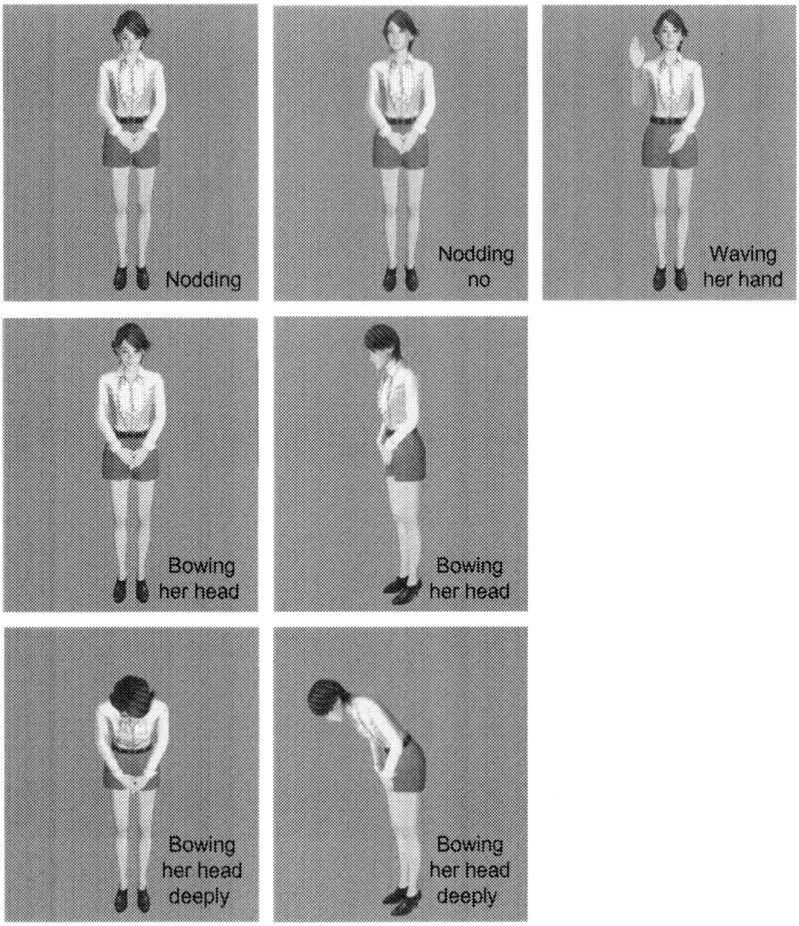

98 Emotions, Technology, and Design

Figure 5.6 Gestures of the system.

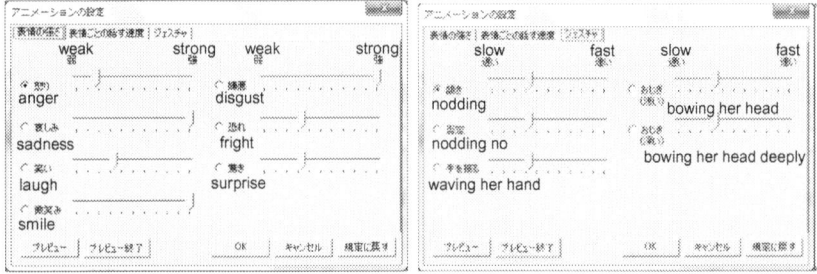

Figure 5.7 Dialog boxes of the facial expression/gestures.

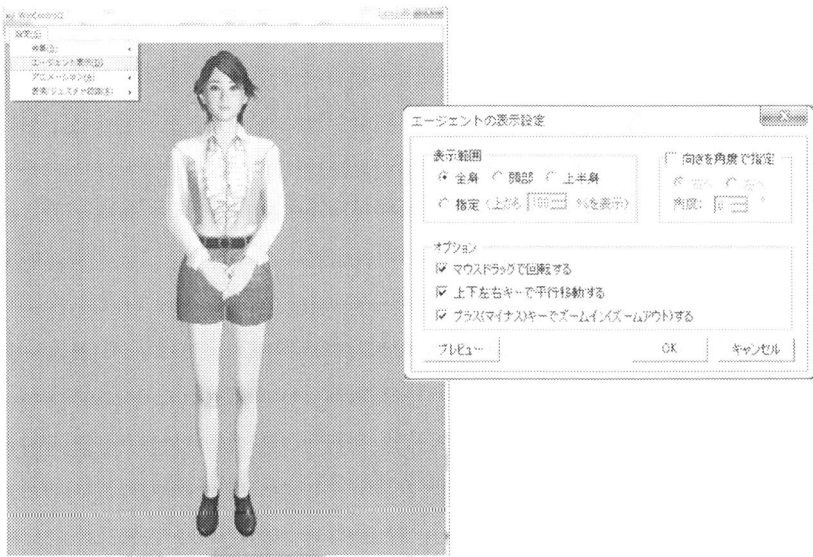

Figure 5.8 A dialog box of the agent display.

Figure 5.9 A spoken agent system for mental care.

(2012), which is a dictionary for translating negative words to positive words. Such counseling is provided in the field of positive psychology. For example, such as follows is talked.

The agent: How was your day? The user: I went to my university.
The agent: What kind of things happened?
The user: I get *frustrated* because only a friend might be good.
The agent: I felt sorry for you. (Facial expression is "Sad") I think feeling frustrated is a source of personal growth. (Positive words from Negapo Dic.)

DISCUSSION

Even if a partner is not a human being, we are perceived by an agent with superficial information such as facial expressions or words. Media Equation (Reeves & Nass, 1996) and Persuasive Technology (Fogg, 2003) researches reveal the same.

From the result of the first experiment in section "Experiment Investigating Impressions and Behavior Change Caused by Replies from the Agent," facial expressions and words are very important because people are easy to be persuaded when the agent had a favorable impression. It is thought that an agent can make in-depth contact with the user by foreseeing the user's emotion and empathizing with the user. From the result of the next experiment in section "Comparative Experiment on Effects of a Virtual Agent and a Robot" it is thought that the agent should have rich emotional expression and should not encourage the user more than necessary. According to the results, we developed a spoken agent system.

Overall, for the experiment of the first version of the spoken agent system, the answers to the questionnaire revealed many positive opinions. Considering the free answers to each question, subjects felt the brain-wave measuring equipment and the speech recognition and synthesis system were novelties, and the system's specialty was that its use was facilitated through dialogs and it was highly motivating. The brain-wave measuring equipment and the speech recognition and synthesis system used in our system were highly motivating; however, the subjects were divided over their use, and some felt them to be awkward to use. Some of them made comments on the speech recognition system such as "It was difficult to recognize words," "It was difficult," etc. Consideration should be given to improving the system's ease of use by developing or using a speech recognition system of greater accuracy. Also, regarding the brain-wave measuring equipment, some made comments on it such as "It was hard to wear" and "It was painful," so it could be considered that it was not easy to accept. Consideration should be given to improving the human interface, including using image processing and the headset concurrently.

The results of the experiment revealed that we must improve the interface of the system, especially for the attachment of the brain-wave measuring equipment. Then we developed the next version of the system for mental care of hikikomori social withdrawal persons or using KINECT, which recognizes the facial expression/gestures of the user.

I think this system is helpful for mental care of hikikomori social withdrawal persons if it can be used easily at home. In popular lore it is said that meaningless words of encouragement are off limits for depressed clients. It is similar in results of the second experiment described in section "Experiment Investigating Impressions and Behavior Change Caused by Replies from the Agent." In addition, because even hikikomori or social withdrawal persons are at home and stay in front of a computer most of the time. I think that supporting these people by a personal computer is a key to solution.

Study of a spoken dialog agent system needs to be investigated further including emotions, technology, and design domains. Our research group is going to advance our own studies of a spoken agent system using superficial information in the future. We will install this spoken agent in the counselor's office of the university, and start to validate the system's effectiveness in the near future.

CONCLUSION

This chapter introduces two experiments and the spoken agent system developed according to the results. We evaluated the first version of the system and revised. Even if a partner is not a human being, we are perceived by an agent with superficial information such as facial expressions or words. Our research group is going to advance our own studies of a spoken agent system using superficial information in the future.

ACKNOWLEDGMENTS

I would like to thank Mizue Nagata for experiments, and Ryuji Ebata for research and technical assistance. This work was supported in part by JSPS KAKENHI Grant-in-Aid for Scientific Research on Innovative Areas Numbers 22118503.

REFERENCES

Daibo, I. (1978). Sansya kan communication ni okeru taijininsyo to genngo katudousei. *jikkenn shinri gaku kenkyu*, 18, 21–34 (in Japanese).
Fogg, B. J. (2003). *Persuasive technology—Using computers to change what we think and do*. Elsevier.
Negapo Dictionary (2012) Syufuno-Tomo-Sya (In Japanese).
Peterson, C. (2006). *A primer in positive psychology* (1st ed.). New York: Oxford University Press.
Reeves, B., & Nass, C. (1996). *The media equation: How people treat computers, television, and new media like real people and places*. New York: Cambridge University Press.
Rizzo, A., Sagae, K., Forbell, E., Kim, J., Lange, B., Buckwalter, J. G., et al. (2011). SimCoach: An intelligent virtual human system for providing healthcare information and support. In *The Inter Service/Industry Training, Simulation & Education Conference (I/ITSEC), Orlando, Florida*.
Seligman, M. E. P. (2004). *Using the new positive psychology to realize your potential for lasting fulfillment*. Atria Books.
Sumi, K. (2012). Human interface of robots or agents via facial and word expression. In *International Symposium on artificial life and robotics (AROB), Oita, Japan*.
Sumi, K., & Ebata, R. (2013a). A character agent system for promoting service-minded communication. *Intelligent virtual agents. Lecture notes in computer science*, LNAI8108 (438 pp.). New York: Springer.
Sumi, K., & Ebata, R. (2013b). Human agent interaction for learning service-minded communication. In *iHAI2013, 1st international conference on human-agent, interaction, 8.2013*.
Sumi, K., & Nagata, M. (2010). Evaluating a virtual agent as persuasive technology. In J. Csapó & A. Magyar (Eds.), *Psychology of persuasion*. Shimane, Japan: Nova Science.
Sumi, K., & Nagata, M. (2012). Characteristics of robots and virtual agents in regard to persuasion to maintain motivation. In *APCHI 2012, 10th Asia Pacific conference on computer human, interaction, 9.2012*.
Wada, K., & Shibata, T. (2007). Robot therapy in a care house—Change of relationship among the residents and seal robot during a 2-month long study. In *Robot and human interactive communication, RO-MAN 2007. The 16th IEEE international symposium, National Institute of Advanced Industrial Science and Technology, Tsukuba* (pp. 107–112).
Weizenbaum, J. (1966). ELIZA—A computer program for the study of natural language communication between man and machine. *Communications of the ACM*, 9(1), 36–45.

CHAPTER 6

Engaging Learners Through Rational Design of Multisensory Effects

Debbie Denise Reese[a], Dianne T.V. Pawluk[b], Curtis R. Taylor[c]
[a]Zone Proxima, LLC, Wheeling, WV, USA
[b]Virginia Commonwealth University, Richmond, VA, USA
[c]University of Florida, Gainesville, FL, USA

Well-designed digital instructional games are powerful learning environments because they guide and reinforce learners to discover and apply targeted to-be-learned knowledge (Gee, 2008, 2009). People naturally construct concepts and categories through sensory transactions with their environment (Jonassen, 2006; Lakoff, 1987). If a game environment's multisensory representations are defined and engineered as analogs of a targeted relational structure, and if instructional game goals guide and reinforce learners to discover and apply just that relational structure, then these multisensory analogs may more effectively increase learning outcomes than unisensory analogs or realistic multimodal effects.

"It is likely that the human brain has evolved to develop, learn, and operate optimally in multisensory environments" (Shams & Seitz, 2008, p. 411). Thus, effectively engineered multisensory instructional environments should enhance learning effectiveness, and multisensory representations could be designed to enhance the effectiveness of instructional digital games. Technically, multisensory refers to stimuli with more than one integrated modality; e.g., the auditory-visual feedback (click and flash) when logging into a touch-screen interface (a stimuli set). Multimodality is a more general term. Multimodal refers to sensory stimuli through more than one modality, but does not require multimodality integration within any one stimuli set. In this chapter, effectiveness refers specifically to learner acquisition (construction) of new knowledge structures targeted by learning outcomes.

Multimodal interactive environments currently exist; however, their success in producing targeted learning outcomes has been problematic.

Many of these environments focus on realism. Unfortunately, real environments provide so much sensory input that it is difficult for novices to focus on the salient information. Also, when environments are realistically portrayed, significant differences can be small compared to the scale of targeted phenomena. Humans are poor at detecting small relative differences across all modalities (Wolfe et al., 2011). This impedes the effectiveness of instruction and learning. We suggest another approach, especially for game-based learning. Analogical reasoning is a natural cognitive process through which people learn (Hummel & Holyoak, 1997). We propose to leverage clear, age appropriate analogical reasoning for rational design using highly salient multisensory effects (see Figure 6.1). Rather than engineer multimodal or multisensory stimuli that realistically represent targeted knowledge, a rational design approach bases engineering on theories explaining how people naturally construct concepts through lived experience and analogical reasoning (Reese, 2009, 2014, 2015b).

As mentioned above, previous work, albeit with realistic simulations, has shown that intuition in multisensory design, such as using force feedback to present the concepts of force (Moore, Williams, Luo, & Karadogan, 2012; Park et al., 2010) may produce counterintuitive, poor results. This may well be due to humans being poor at detecting changes in specifics, including forces, across all modalities (Wolfe et al., 2011). However, this reasoning is not always obvious after completion of the design, let alone a priori, during the design itself. Designers require a sound set of empirically supported guidelines *before* building multimodal learning environments and devices. In this chapter, we present foundational underpinnings that might provide a viable rational approach toward design of multisensory, game-based instructional systems.

The rational approach defines unique modality representations as analogs of targeted knowledge. It leverages the cognitive processes underlying

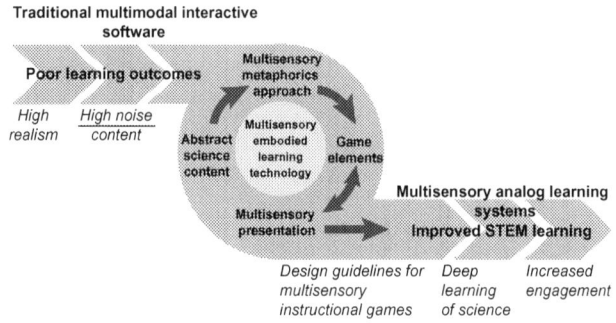

Figure 6.1 Hypothesized impact of multisensory analog learning systems.

concept formation by mapping defined modal representations to embodied [virtual] transactions that model targeted knowledge structures. We also propose to use innate modality-specific perceptual processing mechanisms, such as selective attention and grouping, to unconsciously reinforce the analogy relationships and attentional flow of the learning process (Goldstone, Landy, & Son, 2010). We propose application of this rational approach to produce game modules that act as learning systems to guide and reinforce learners to discover and apply targeted knowledge through those transactions. We also propose that scientists manipulate those game modules for testing hypotheses that identify and validate design principles for analogs when represented in haptic and auditory, as well as visual, modalities.

DRAWING ON TWO PROCESS MODELS FOR REPRESENTATION

Analogical Reasoning

Analogical reasoning theory provides a foundation for metaphor-enhanced design of multisensory representations. Analogists have established that people learn from mapping relational structure from one domain to another (Gentner, 1983; Gentner & Markman, 1997; Holyoak, Gentner, & Kokinov, 2001). Figure 6.2 displays the CyGaMEs approach (Reese, 2007, 2009) to the application of cognitive science analogical reasoning theory in the design of digital game instructional systems. The model begins with the rectangle on the left representing the real-world relational structure for some targeted knowledge domain. The ovals signify objects (e.g., people, places, events, things) and the directional arcs represent the relations connecting the objects. When people encounter new, unfamiliar, or abstract domains, people make inferences (learn) by mapping relational structure from the known domain (left) to the unknown domain (the right-hand rectangle). Immediate, pragmatic goal structures shape the characteristics of those mappings (Gentner & Holyoak, 1997; Holyoak, 1985, 2012; Holyoak & Thagard, 1989; Spellman & Holyoak, 1996).

In metaphor-enhanced design approaches, the designer uses a specification (knowledge map) of the to-be-learned knowledge domain (the right-hand rectangle) to constrain, guide, and evaluate a game world system for alignment with that knowledge domain (left-hand rectangle). The game world models the relational structure of targeted domain knowledge, and the game goals (analogs of targeted learning goals) guide and reinforce the player to discover and apply targeted new knowledge. Thus, the learner

106 Emotions, Technology, and Design

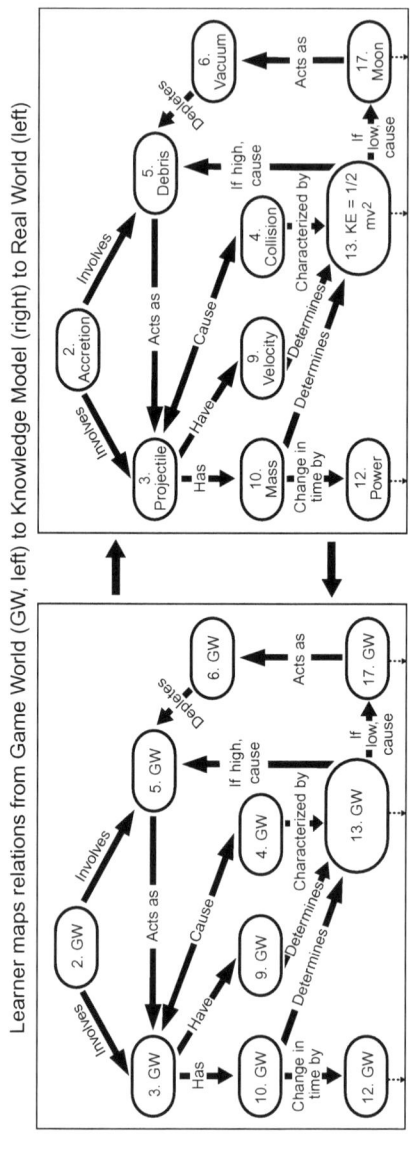

Figure 6.2 The CyGaMEs approach to application of cognitive science analogical reasoning theory to design of instructional digital game systems.

constructs a mental model (right-hand rectangle) that represents targeted, real-world knowledge (e.g., scientific phenomena, the left-hand rectangle).

Procedurally, engineering multisensory metaphor enhanced game-based learning environments begins with domain specification. In collaboration with subject matter experts, a learning engineer who specializes in this technique (metaphorist) conducts a cognitive task analysis (Chipman, Schraagen, & Shalin, 2000; Clark, Feldon, Merriënboer, Yates, & Early, 2008) to discover and specify the relational structure of the to-be-learned domain. Then multisensory human-computer interface designers, game designers, and game developers invent (define) and realize each sensory representation as an integrated and integral component of the embodied learning system (game world: game goals, game rules, mechanics, feedback, etc.). In other words, the game design team translates targeted knowledge-domain relationships into player transactions with the game world. In a multisensory learning game, these transactions might derive from haptic or auditory sensory representations in addition to or in place of the more typical visual stimuli.

Transactions may also take advantage of innate sensory-specific processing mechanisms to facilitate the mappings, at least initially. This is because perceptual factors such as Gestalt principles provide a grounding that influences symbolic reasoning (Goldstone et al., 2010). For example, Goldstone et al. studied the influences of perceptual grouping on symbolic reasoning in arithmetic. According to conventions for order of precedence of operations in arithmetic calculations, multiplication should occur before calculations using addition. Goldstone et al. applied the Gestalt principle of visual proximity to facilitate students' application of the order of operations. When multiplication symbols and numbers were placed closer together and those for addition were placed farther apart, the students processed equations involving order of operations more quickly. When spatial relationships were reversed, the processing of equations occurred more slowly and the error rate increased (six times as many errors when spacing was inconsistent with the operator convention). Even though participants received immediate feedback after each trial, behaviors were robust. Goldstone et al. concluded that "perhaps all cognition may intrinsically involve perceptually grounded processes" (p. 31) and "grounding cannot be ignored for any cognitive task" (p. 27). Scaffolds using perceptual grounding, like proximity, can be faded and removed after learning. Indeed, Goldstone et al. suggest fading will promote transfer from a specific instructional context to more general application.

An Integrated Process for Design and Assessment

The CyGaMEs approach structures a Metaphorics process for translating real-world scientific phenomena into game worlds that teach as well as assess learning. Metaphorics is a set of steps that (a) apply targeted theory from the learning and cognitive sciences, (b) specify knowledge through cognitive task analysis, (c) embody knowledge as a transactional digital game, (d) measure knowledge through embedded assessment, and (e) integrate knowledge building into instruction. CyGaMEs Metaphorics is consistent with the knowledge-learning-instruction framework (KLI; Koedinger, Corbett, & Perfetti, 2012). Figure 6.3 integrates CyGaMEs Metaphorics within KLI to explain the connection between targeted knowledge, the CyGaMEs game world, and CyGaMEs game-based assessment. Scientists build a consensus model of the relational structure of targeted natural phenomena (expert knowledge components). The learning scientist conducts cognitive task analysis to specify a knowledge map or ontology (knowledge component representations). A metaphorist guides the game design team to translate that map into an analogous game system of instructional events consisting of game transactions (transactional knowledge components). The game is constructed to produce metrics of player progress toward each game goal (measured knowledge components). From those assessment events and measured knowledge component representations, the existence of the player's (learner's) knowledge components (mental models) and the

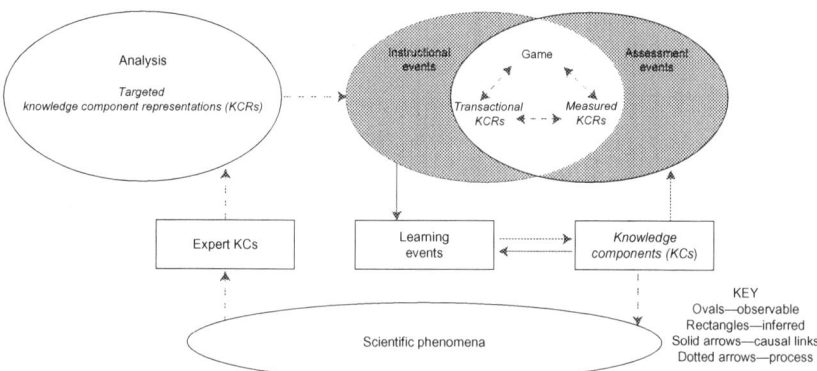

Figure 6.3 Integration of the CyGaMEs approach (e.g., Reese, 2009; Reese et al., 2013) with the knowledge-learning-integration framework. *Copyright 2014 by Debbie Denise Reese. Used with permission. Adapted from Koedinger et al. (2012). Copyright 2012 by the Cognitive Science Society, Inc. Adapted with permission.*

learning events that caused them are inferred. In this way such instructional games both cause and measure (assess) learning.

Alignment among the to-be-learned knowledge, gameplay, and embedded assessment is essential for instructional games that cause and measure learning (Mislevy, Behrens, Dicerbo, Frezzo, & West, 2012; National Research Council, 2011). CyGaMEs Metaphorics guides alignment during design and development and provides methods and metrics to quantify degree of compliance during evaluation.

Beyond Realism: Cognitive Load Considerations

Noise, Attention, and Prior Knowledge

Many implementations of educational software in science areas try to mimic real environments (eTouchSciences, LLC; eTouchSciences, 2014a,b; Rose, 2012; TechConnectWV, 2014). Consideration of what is known about sensory perception reveals two inherent weaknesses to this approach. First, there is a potential to inundate the learner with excessive noise that hinders perception of what is to be learned (Wolfe et al., 2011): the background (needed for realism) in which the focus content is provided obscures what is to be learned. This can be related to providing too much extraneous cognitive load (Paas, Renkl, & Sweller, 2003; Sweller, van Merrienboer, & Paas, 1998). A limited-capacity working memory requires people to selectively attend to stimuli. Novices lack knowledge necessary for filtering extraneous information. Second, perceptual tasks in the real world, even as simple as searching for an object, rely heavily on a priori knowledge (Wolfe et al., 2011) and can otherwise result in feelings of frustration for the learner. This is knowledge that novices do not have in a new domain. A rational approach would engineer, test, and produce effective multisensory metaphors for a given age group, country, etc. for incorporation into an instructional digital game. The game would scaffold players' discovery and application of those metaphors to support clear focus on the core concepts to be learned, and possibly result in pleasure and encouragement.

Cognitive Load Theory

The rational design of multisensory game-based instructional systems is supported by cognitive load theory (Hollender, Hofmann, Deneke, & Schmitz, 2010; Paas et al., 2003; Sweller et al., 1998). Cognitive load theory identifies the conscious processes of thinking as working memory. As Newell and Simon explained,

> *Awareness here means that the subject's immediately subsequent behavior can be a function of the given information [in working memory]. It does not mean that the subject can report that he has the information. For example, he may be too absorbed in the task to be "aware" in this monitoring sense.*
>
> *(Newell & Simon, 1972, p. 804)*

Working memory has limited capacity, with a maximum duration of about 20 s (Paas & Sweller, 2012), ability to hold (e.g., recall a list of digits) about seven chunks of information (Miller, 1956), and with a maximum concurrent processing limit of two to four chunks of information (Paas & Sweller, 2012; Sweller et al., 1998). Processing activities would include "organizing, contrasting, comparing," and other manners of working on information (Sweller et al., 1998, p. 252) such as problem-solving (Paas & Sweller, 2012). The key process, and connection to the CyGaMEs approach, is captured by the word "chunk."

Cognitive load theory identifies three types of load. *Intrinsic* cognitive load is indigenous to the to-be-learned information and task. Unnecessary information or activities, such as noise or nongermane activity or stimuli, add *extraneous* cognitive load. Nongermane instructional activities are defined as those that do not scaffold the learner to construct relevant and viable schemata. *Germane* cognitive load scaffolds the learner to construct viable and relevant schemata.

Expanding Cognitive Capacity for Learning Through Chunking and Schemata

Chunking unlocks human capacity to exceed initial limits of working memory (Chi, Glaser, & Rees, 1982; Miller, 1956; Newell, 1990; Newell & Simon, 1972). Chunking is the key to learning and acquisition of expertise. Miller (1956) provided a concrete example of how chunking could extend the capacity of working memory for digits. Recalling a sequence of 18 binary digits such as 1 0 1 0 0 0 1 0 0 1 1 1 0 0 1 1 1 0 is beyond typical working memory capacity. With practice and recoding, an individual can chunk the 18 binaries into a much more manageable six chunks. First, the individual binaries are grouped into six octal digits: 1 0 1, 0 0 0, 1 0 0, 1 1 1, 0 0 1, 1 1 0. Using a recoding scheme for translating each octal triplet into a decimal-digit name (000=0, 001=1, 010=2, 011=3, 100=4, 101=5, 110=6, 111=7), the 18 binary digits are recoded as six digits (5 0 4 7 1 6). Miller reported that using a similar recoding system, Sidney Smith was able to recall and repeat 40 binary digits without error. The nature of the CyGaMEs approach naturally chunks information.

The player builds knowledge structure chunks through integrated gameplay transactions.

Newell and Simon's (1972) theories of human problem solving and cognition (Newell, 1990) built on Miller's chunking research and de Groot's (1946/1978) seminal finding that master chess players' superior ability to reconstruct chess board configurations was confined to meaningful configurations. Within long-term memory, people chunk information as meaningful categories. The knowledge associated with a category may be referred to as a schema (Chi et al., 1982). As relevant knowledge and experience expand, information chunks are subsumed into higher level chunks (e.g., for explanation of how this occurs in verbal learning, see Ausubel, 1962, 1963). When analogists speak of profound or deep relational structure, they mean that a domain of knowledge contains a highly differentiated hierarchical structure of levels and branches. Due to chunking and hierarchical knowledge structures, a chess master might recognize the exact game from which a key chessboard layout derives, encapsulating play strategies and devices. Research across knowledge domains supports chunking and subsumption involved with problem solving and learning. For example, physicists conceptualize and approach physics problems according to higher-order principles (Chi et al., 1982). Skilled electronics technicians chunk by their knowledge of "relationships among entire functional units" (Egan & Schwartz, 1979, p. 156). By design, CyGaMEs naturally facilitates the subsumption of chunks into higher-level chunks and schemata formation.

Chunking with densely hierarchical relational structure is recognized as a fundamental aspect differentiating expert from novice knowledge (Bransford, Brown, & Cocking, 2000). Novices to a domain tend to think, reason, and analogize according to superficial features knowledge systems. There is long-standing consensus that a primary role of instruction is to develop learning environments that guide and reinforce learners to construct viable, integrated, coherent, and robust knowledge structures, although pedagogical approaches may diverge (for example, compare Linn, Davis, & Bell, 2004; Sweller, Kirschner, & Clark, 2007).

The nature of the CyGaMEs approach naturally chunks information to build up the knowledge schemata through gameplay transactions. Instructionally, CyGaMEs seeks to obviate superficial knowledge construction through knowledge integration. Knowledge integration is a matter of nesting new knowledge within existing cognitive structures. The core aspect of analogical reasoning illustrated in Figure 6.2 is a ubiquitous cognitive process (Hummel & Holyoak, 1997) that enables learning and higher order reasoning by a mapping of relational structure from a source domain that is concrete or relatively familiar to a target domain that is abstract or relatively unfamiliar (Holyoak, 2012; Holyoak & Thagard, 1997; Polya, 1954). The systematicity principle (Gentner, 1983) explains that people prefer mappings of deep relational structure when they have sufficient experience and prior knowledge. Indeed, it is deep relational mappings from source to target domains that support construction of deeply integrated cognitive models. However, children and novices (like inexpert chess players or electronics technicians) will make mappings according to superficial similarities because they lack sufficient experience and prior knowledge. Superficial mappings may be used for memorization, but they do not incur relational understanding that promotes or comprises growing expertise, and they do not accurately support higher order cognitive activities such as problem-solving. A CyGaME's learning environment guides and rewards learners to construct appropriate prior knowledge, on which to scaffold new concepts. Thus, CyGaMEs extends the capability of novices (and children) to chunk and construct the schemata that comprise systematicity. Instructional design principles support knowledge construction through activation or production of apt prior knowledge (Merrill, 2002). CyGaMEs learning environments prepare learners with viable knowledge structures (prior knowledge) foundational to deep domain knowledge. Knowledge structures function as chunks. Such instructional interventions prepare learners for knowledge acquisition; they prepare learners for future learning (Reese, 2007).

As demonstrated by studies of chess masters and digit span memory, schemas from long-term memory can act as a chunk within working memory. They enhance capacity while reducing cognitive load. When schemata are learned to the point of automaticity, demand on working memory capacity is greatly reduced (Paas et al., 2003; Sweller et al., 1998) and "largely by-passed" (Sweller et al., 1998, p. 256). Indeed,

because of schema construction, although there are limits on the number of elements that can be processed by working memory, there are no apparent limits on the amount of information that can be processed. A schema, consisting of a single element in working memory has no limits on its informational complexity. In summary, schema construction has two functions: the storage and organization of information in long-term memory and a reduction of working memory load. It can be argued that these two functions should constitute the primary role of education and training systems.

(Sweller et al., 1998, p. 256, italics added)

The degree of systematicity, what Paas and their colleagues describe as interactivity between relevant elements (Paas et al., 2003), determines intrinsic load. In general, increasing systematicity of to-be-learned knowledge and skills increases cognitive load. However, chunking reduces intrinsic load because complexity may be reduced to one subsuming element. By design, a CyGaMEs environment guides and reinforces the learner to build viable relational structures (schemata). The repetitive nature of such gameplay, in which each gesture is a new problem-solving challenge inquiry, allows the learner to incorporate targeted schemata with an embodiment that approaches automaticity, although such knowledge may be tacit (i.e., preconceptual, see Hatano & Inagaki, 1986 for definition; for description of application within CyGaMEs processes, see Reese, 2007, 2012).

Expanding Capacity Through Multisensory Stimuli

Multisensory presentations can reduce cognitive load (Shams & Seitz, 2008; Sweller et al., 1998; van Gerven, Paas, van Merriënboer, & Schmidt, 2006). This is because each sensory system is posited to have its own working memory, which can work simultaneously in parallel with the others (Sanman & Stanney, 2006). To address information processing for each sensory system, one approach to facilitate learning would be to apply the concepts from visual human computer interaction (HCI), such as cognitive load split-attention, modality, and redundancy effects (Paas et al., 2003), to all sensory systems (Hollender et al., 2010). However, the processing capabilities of different sensory systems vary widely. For example—in vision, color, orientation, size, and several other features can be searched across multiple items in parallel (Wolfe & Horowitz, 2004). In contrast, for touch material properties (e.g., roughness, compliance, and thermal properties) can be searched in parallel across items, while geometric information (e.g., orientation) requires significant cognitive processing (Lederman & Klatzky, 1997). For audition, segmentation is described in terms of separating audio streams, where it is quite normal to process more than one stream simultaneously with ease (Wolfe et al., 2011). Together, these examples suggest that different categories of metaphors may map more effectively onto one or another sensory system based on the use of the most effective features, and the most efficient processing methods for each system. Pawluk (Burch & Pawluk, 2011) has used some of these ideas to effectively design new displays for individuals who are blind or visually impaired that have resulted in significantly improved performance. In addition, although cross-modality interaction effects are known to occur in perception, little is known how they affect learning.

Facilitation Through Innate Perceptual Mechanisms

Innate principles of perceptual organization in the different sensory domains can scaffold learner transactions with multisensory metaphors to aptly discover, define, and apply targeted knowledge. Each sensory system has a separate, finite amount of "perceptual memory." Multisensory characteristics such as attention and grouping can be engineered to produce an automatic perceptual relationship between appropriately sensory-represented cognitive concepts (multisensory metaphors). For example, people perceptually organize things that are closer together differently than things that are further apart. These perceptual organization elements are preattentive. Preattentive elements are separate from game analogs, which are representations of the concept. Preattentive elements would complement relational structure represented by the modality metaphor and facilitate a natural or automatic making of connections. For example, players naturally process relationships perceptually, based upon proximity. So spatial clusters of connected game elements representations would capitalize on natural perception. *Perceptual organization is a method for controlling the flow of thought processes that places relatively low demands on cognitive load*. This hypothesis is supported by previous research. For example, Wallace, West, Ware, and Dansereau (2010) used gestalt principles (color, shape, and proximity) to organize information within concept maps. Participants who processed information from these enhanced maps recalled statistically significant greater amounts of information than did those who learned from text narrative or from unenhanced maps. Within learning games, this gestalt organization is both spatial and temporal.

Some features of an object are processed more quickly and are more likely to be used than others. They guide subsequent attention even if multiple features are used (Wolfe & Horowitz, 2004). These preattentive object features and grouping principles have been described for vision and haptics in the sensory perception literature (Lederman & Klatzky, 1997; Overliet, Krampe, & Wagemans, 2012; Wolfe & Horowitz, 2004; Wolfe et al., 2011) and are summarized in Table 6.1. Although preattentive features have not been identified in audition, basic parameters for sonification (the use of nonspeech sounds) and similar ones for tactile vibrations have been developed (see Table 6.2), along with the determination of grouping principles for those modalities (Brewster & Brown, 2004; Denham & Winkler,

Table 6.1 Preattentive Vision and Haptics Dimensions (Features for Guiding Attention) and Grouping (Association) Principles

	Dimensions: Features for Guiding Attention	Grouping Principles
Vision	• Color • Motion • Orientation • Size (including length and spatial frequency) • Perhaps more	• Similarity • Proximity • Common Fate (generalization to similarity in changes) • Synchrony (changes at the same time)
Haptics perception[a]	• Roughness • Stiffness • Stickiness (friction on skin) • Temperature	• Similarity • Proximity • Common Fate and Synchrony (require simultaneous contact with objects)

[a] Commonly refers to the exploration of objects (e.g., both the sense of touch and proprioception-which uses the receptors in the joints and muscles).

Table 6.2 Audition and Tactile Dimensions (Parameters) and Association (Grouping) Principles

	Basic Dimensions (Parameters)	Grouping Principles
Audition	• Magnitude • Pitch • Timbre • Order • Rhythm • Attack and Decay	• Spatial adjacency (3-D) • Spectral adjacency • Temporal adjacency • Timbre • Onset
Tactile sensation[a] (Vibration)	• Magnitude • Pitch • Rhythm • Attack and Decay	• Not known but potentially similar to audition

[a] Commonly refers to the skin's sensitivity (e.g., the response on the fingertips to vibration).

2013; Schertenleib & Candey, 2013). The proposed research program should also consider and investigate cross-modality interactions. Cross-modality interactions have been found to direct overall spatial attention (Driver & Spence, 1998; Wesslein, Spence, & Frings, 2014).

Our approach to multisensory presentation is novel in that it considers highly salient preattentive object features and grouping principles in three sensory systems (vision, audition, and haptics/touch) for implicitly guiding attention and presenting metaphors effectively to facilitate learning. Although multisensory HCI interfaces are in existence, most do not consider haptics/touch as well as vision and audition, nor do these interfaces consider the use of "preattentive" object features, grouping principles have only been used for static visual design, and no previous system to our knowledge has considered perceptual aspects of attention and grouping processes to facilitate the flow of learning.

Section Summary

We have come to the end of our presentation of the theory underlying a rational approach to multisensory effects in digital instruction game design. Analogical mapping (Figure 6.2) and the CyGaMEs elaboration of the Knowledge-Learning-Instruction framework (Figure 6.3) provide theoretical rationale and process steps for domain specification (knowledge mapping through cognitive task analysis) and embodiment of targeted knowledge as a transactional game system. Game goal analogs of targeted learning goals guide and reward players' transactions as they discover and apply targeted knowledge. The design team will define multisensory effects that embody components of the target domain relational structure. The use of multisensory systems is based on cognitive load theory (Samman & Stanney, 2006), whereas the actual design of multisensory effects is based on current knowledge of human perception. Cognitive load theory posits each sensory system has its own short-term memory. Use of multiple sensory systems increases the amount of information that can be processed (Hollender et al., 2010). All senses have exhibited properties that facilitate grouping of items in the environment (see Tables 6.1 and 6.2). However, the grouping properties are different between the senses and not completely known in haptics/touch (Overliet et al., 2012; Wolfe et al., 2011).

TOWARD RATIONAL DESIGN GUIDELINES

Earlier in this chapter, we provided a foundation for design, research, and development investigating how metaphor-enhanced multisensory representations interact with multisensory dimensions to facilitate learning of abstract science. To develop guidelines, game systems could be designed to systematically examine and determine which multisensory parameters presented in which ways during embodied learning gameplay, are most effective in producing discovery, application, and transfer of knowledge. Results would produce concrete, evidence-based design guidelines and procedures for developing the multisensory representations and possible interactions. By design, such a multisensory system would support knowledge construction. Support of a multisensory system could be accomplished by key confluent elements in the learning environment including embodied instructional gameplay, embedded analytics assessment and adaptability, modular component design, and integration of attentional design constructs involving vision, audition, and haptics.

Using Visual, Auditory, and Haptic/Tactile Multisensory Stimuli

People are familiar with auditory and visual stimuli. Haptics commonly refers to the exploration of objects through both the sense of touch and proprioception (using receptors in the joints and muscles). Tactile commonly refers to the skin's sensitivity (e.g., the response on the fingertips to vibration). We will focus on the perception and manipulation of vibration as it is the easiest dimension to control and is well understood. Very little is known about the interaction of these sensory systems between their dimensions (e.g., color and texture) and on learning or on affect during learning. Interactions between auditory and tactile vibrations may be problematic in some dimensions. Other sensory systems are very dissimilar on their most salient dimensions; however, other potential cross-modal interactions need to be considered.

In addition, although previous work by Reese incorporated some sonification within a CyGaMEs learning environment (CyGaMEs, 2013), her research designs never formally controlled for sonification parameters. It is important to investigate high saliency features and grouping effects of the visual, auditory, and/or haptic/tactile systems on learning and affect through experimental control of parameter mapping. Using an iterative design process, a research team can determine the number of values along a feature dimension that one is able to effectively identify. This information is an important part of the proposed guidelines. Although the best discrimination abilities under controlled conditions have been quantified in the sensory perception literature, the benefit of using these features only applies when the discrimination is easily performed. The iterative design process enables testing and refinement of assumptions about which, where, when, how, and for whom these modalities are effectively combined to cause learning.

Although a variety of affordable commercial methods can be used to produce visual and auditory effects, haptics and tactile effects are much less standard. The commercially available Novint® Falcon is an affordable device that can produce force and sense position along the x, y, z axes. For tactile vibration, Burch and Pawluk (2011) have designed low cost custom-made tactile vibrators for the fingertips using piezoelectric speakers. The Novint Falcon can produce haptic effects and the tactile vibrators can produce tactile effects. Pawluk has previously created most of these effects in her laboratory experiments. Her algorithms take into account perceptual versus physical dimensions and new effects not previously created by others. Burch and Pawluk (2011) have found that using uniquely identifiable values for tactile pseudo-textures effectively and efficiently supports discrimination of values along feature dimensions.

Learning and Embedded Feedback/Assessment

Integrated design methods like CyGaMEs can be used to investigate the use of high saliency ("preattentive") features and grouping effects of the visual, auditory, and/or the haptic/tactile systems individually or in combination to relay metaphors effectively. As illustrated within Figure 6.3, methods like CyGaMEs produce instructional digital game systems that simultaneously function as learning events and assessment events. When a game goal is the analog of targeted learning goals, and the game system analog guides the learner to discover and apply targeted relational structure, instructional game transactions can be engineered to measure player progress toward each of those goals (Reese et al., 2012; Reese, Tabachnick, & Kosko, 2013).

CyGaMEs designs a Timed Report that measures and posts player progress every 10 s of gameplay. Timed Reports posts with paradata which contextualize each observation with relevant, concurrent information (e.g., time stamp, location in the environment, goal at time of collection). In a CyGaMEs environment, each player gesture is one meaningful player transaction that has the potential to change the game state. The Timed Report analyzes the result (goal attainment) of a player's gameplay gestures at the end of each 10-s interval. This time interval is chosen as, in his powers of 10 analysis of the time scale of human action, Newell (1990, p. 149) identified the 10-s interval as

- The time frame within human cognition "where knowledge is available from the environment about what to learn, namely experience attained in working at $\sim$10 s,"
- Governed by "the opportunity for acquisition,"
- Locating its "cause in conditions for learning," and
- Working "according to representational law ... [to] represent things—objects, relations, and activities in the external world."

Newell's conceptualization further aligns with CyGaMEs Metaphorics, as Newell recognized that cognitive task analysis could be used to specify cognitive activity at this level.

Timed Report data and analyses (Reese, 2015a; Reese et al., 2012, 2013) have identified learning trajectories and individual and aggregate learning levels, rate of learning, and acceleration (any changes in rate such as aha! or learning moments). Metrics such as the Timed Report permit observation and analysis of the causal effects of novel multisensory game metaphors and high saliency features and grouping effects of the visual, auditory, and/or haptic/tactile systems on learning.

The Timed Report is an embedded assessment measure. Such game-based assessments are characterized as stealth assessment (Shute, 2011; Shute & Ventura, 2013; Shute, Ventura, Bauer, & Zapata-Rivera, 2009) because the gameplay is the assessment (see Figure 6.3). The game monitors the player continuously, in the background. The player is unaware of the assessment and the assessment does not interrupt the player or gameplay.

Measuring Affect

The CyGaMEs Flowometer Report is a measure of affect: a player's immediate subjective perceptions of current experience (see Figure 6.4). Although the Flowometer is integrated into a game system, it is not an embedded assessment because the Flowometer stops gameplay and asks the player to report, on a scale from 0 (low) to 100 (high), perceived level of skill and challenge (Reese, 2010). Reese derived the Flowometer, protocols implemented to administer it, and methodologies for analysis for Flowometer Report data (e.g., Reese, 2015) directly from flow theory and research methods (Csikszentmihalyi & Csikszentmihalyi, 1988b; Csikszentmihalyi & Larson, 1987; Delle Fave, Massimini, & Bassi, 2011; Hektner, Schmidt, & Csikszentmihalyi, 2007; Massimini & Carli, 1988; Moneta, 2012). Skill and challenge dimensions may be analyzed together using multilevel models. Data can be plotted as trajectories of player experience over time (Reese, 2015). CyGaMEs algorithms have been developed to categorize each skill-challenge dyad into sectors (see Figure 6.5). Based upon Flow literature (Csikszentmihalyi & Csikszentmihalyi, 1988a; Delle Fave et al., 2011; Hektner et al., 2007; Massimini & Carli, 1988) but adapted based upon Hatano and Inagaki's (1986) work with adaptive and routine expertise, the nine sectors of affect are flow, arousal, anxiety, worry, apathy, intrinsic motivation, boredom, and routine expertise (replaces relaxation).

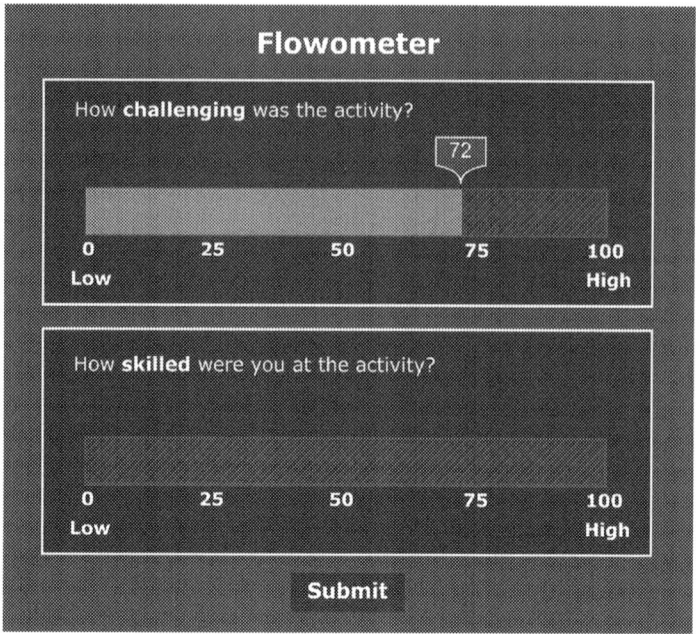

Figure 6.4 The CyGaMEs Flowometer. *Copyright 2008 Debbie Denise Reese and James Coffield. Used with permission.*

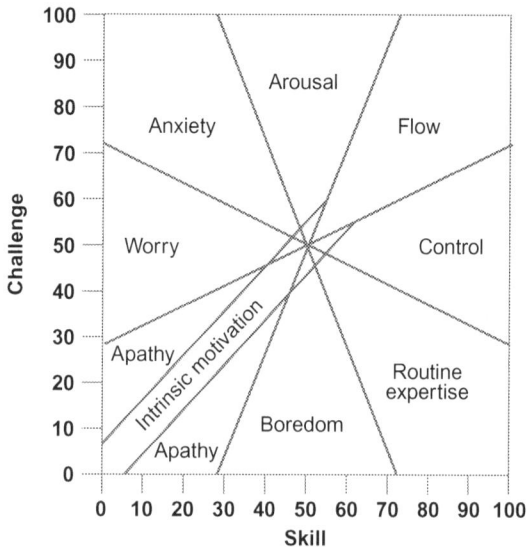

Figure 6.5 The nine-channel model of subjective experience. *Copyright 2014 Debbie Denise Reese. Used with permission.*

CyGaMEs protocols administer the Flowometer within every 5 minutes of gameplay at a randomly determined, preselected time such that each player within a game or nongameplay module receives the Flowometer on the same schedule. Flowometer Reports post with relevant paradata. Contextual paradata such as timestamp and location in the environment are the same variables as those posted with the Timed Report.

When players of the CyGaMEs instructional science game *Selene: A Lunar Construction GaME* actively strive toward expertise, they often perceive gameplay challenge to be greater than their gameplay skill level (Reese, 2015). Players who tenaciously strive toward gameplay achievement—hour after hour—are certainly invested and engaged in their gameplay. But the research methods and measures described find that these players' self-perceptions during gameplay learning do not typically categorize as flow. Rather, gameplay inducing and exhibiting persistence and problem-solving (targeted as twenty-first century skills) may quantify as worry, anxiety, and arousal. Since these players persevere to success, such gameplay experience would categorize as eustress (positive stress during adaptation) rather than distress (negative stress that debilitates); that is, emotions of striving (to learn more about the distress/eustress distinction, see Selye, 1974, 1975, 2013).

What effect, if any, do multisensory representations contribute to subjective self-perceptions of experience during gameplay learning? Integration of flow theory, the Flowometer, and analyses techniques will support exploration of the three-way interaction among multisensory effects, game-based learning, and affect.

Analysis: Toward Design Guidelines

Consistency in data structure creates player logs that contain an integrated record of all game context and player transaction data for nested within player and any other relevant levels of analysis (e.g., classroom). CyGaMEs learning dynamics procedures (Reese et al., 2013) and inferential statistics, case studies, or machine learning can be applied to empirically measure and test the effectiveness of multisensory metaphors, sensory system dimensions, and their interplay with each other, learning, and affect.

INCREASED INTEREST AND ENGAGEMENT

Our proposed rational design applies the CyGaMEs Metaphorics process to engineer and interactively refine multisensory metaphors analogous to targeted knowledge relational structure. The metaphors are realized within game-based virtual environments that include innate sensory processing mechanisms to guide and reinforce players to discover and apply that relational structure through gameplay gestures involving transactions with those multisensory metaphors. If the metaphors and environments are successful and effective, players will construct schemata for targeted knowledge. These emergent knowledge structures provide apt, viable, coherent, integrated cognitive structure on which learners and subsequent instruction can build. They will support deep knowledge structures for targeted learning. The four-stage model of interest development (Hidi & Renninger, 2006; Renninger & Hidi, 2011) attributes growth in interest to a combination of positive valuing, affect, and knowledge. As illustrated in Figure 6.1, we expect that growth of deep knowledge structures would provide the type of knowledge growth that accompanies interest and leads to increased engagement.

THE RESEARCH AGENDA

We have described how rational design applies cognitive science analogical reasoning theory to guide and constrain engineering of multisensory metaphors. The approach supports design and development of game-based learning environments engineered to guide and reward learners to construct knowledge of associated targeted domain learning and skills. Thus, any resultant increase in cognitive load due to effective multisensory metaphors would be germane rather than extraneous. Any increase in cognitive load would be designed as germane because it creates hierarchic schemata construction for deeper learning. This, in turn, supports subsequent chunking characteristic of growing expertise and results in reduced cognitive load during future-related learning. The use of multisensory channels, preattentive features, and other dimensions of sensory elements could enhance player processing of those multisensory metaphors without adding substantially to load on working memory. Embedded measures of learning during gameplay and the Flowometer provide metrics for measuring the effect of these multisensory components on learning, affect, and the interplay between them. This is expected to enable the development of effective design guidelines.

In addition to different sensory systems possibly mapping to different metaphors, the mechanisms by which these sensory systems interact remains unclear, and the implications for multisensory learning environments need to be elucidated. Future research and development should systematically examine and determine which and how multisensory parameters presented during embodied learning gameplay are most effective in producing discovery, application, and transfer of knowledge.

REFERENCES

Ausubel, D. P. (1962). A subsumptive theory of meaningful verbal learning and retention. *Journal of General Psychology, 66,* 213–224.
Ausubel, D. P. (1963). *The psychology of meaningful verbal learning.* New York: Grune & Stratton.
Bransford, J. D., Brown, A. L., & Cocking, R. R. (Eds.), (2000). *How people learn: Brain, mind, experience, and school.* Washington, DC: National Academy Press.
Brewster, S., & Brown, L. M. (2004). Tactons: Structured tactile messages for non-visual information display. In *Paper presented at the 5th Australasian user interface conference, Dunedin, New Zealand.* January 18-22.
Burch, D., & Pawluk, D. (2011). Using multiple contacts with texture-enhanced graphics. In *Paper presented at the World haptics 2011 conference proceedings, Istanbul, Turkey.* June 21-24.
Chi, M. T. H., Glaser, R., & Rees, E. (1982). Expertise in problem solving. In R. J. Sternberg (Ed.), *Advances in the psychology of human intelligence* (vol. 1, pp. 7–75). Hillsdale, NJ: Erlbaum.
Chipman, S. F., Schraagen, J. M., & Shalin, V. L. (2000). Introduction and history. In J. M. Schraagen, S. F. Chipman, & V. L. Shalin (Eds.), *Cognitive task analysis* (pp. 3–23). Mahwah, NJ: Lawrence Erlbaum.
Clark, R. E., Feldon, D. F., Merriënboer, J. J. G., Yates, K. A., & Early, S. (2008). Cognitive task analysis. In J. M. Spector, M. D. Merrill, J. v. Merriënboer, & M. P. Driscoll (Eds.), *Handbook of research on educational communications and technology* (3rd ed., pp. 577–593). New York: Lawrence Erlbaum.
Csikszentmihalyi, M., & Csikszentmihalyi, I. S. (1988a). Introduction to part IV. In M. Csikszentmihalyi & I. S. Csikszentmihalyi (Eds.), *Optimal experience: Psychological studies of flow in consciousness* (pp. 251–265). New York: Cambridge University Press.
Csikszentmihalyi, M., & Csikszentmihalyi, I. S. (Eds.), (1988b). *Optimal experience: Psychological studies of flow in consciousness.* New York: Cambridge University Press.
Csikszentmihalyi, M., & Larson, R. (1987). Validity and reliability of the experience-sampling method. *The Journal of Nervous and Mental Disease, 175*(9), 526–536.
CyGaMEs. (2013). *Selene.* Retrieved July 2, 2012, from http://selene.cet.edu/.
de Groot, A. D. (1978). *Thought and choice in chess.* New York: Amsterdam University Press. Retrieved from, http://books.google.com.au/books?id=b2G1CRfNqFYC&printsec =frontcover&source=gbs_ge_summary_r&cad=0#v=onepage&q&f=false (Original work published 1946).
Delle Fave, A., Massimini, F., & Bassi, M. (2011). *Psychological selection and optimal experience across cultures: Social empowerment through personal growth.* New York: Springer.
Denham, S. L., & Winkler, I. (2013). Auditory perception organization. In J. Wagemans (Ed.), *Oxford handbook of perceptual organization.* Oxford, UK: Oxford University Press.
Driver, J., & Spence, C. (1998). Crossmodal attention. *Current Opinions in Neurobiology, 8,* 245–253.
Egan, D. E., & Schwartz, B. J. (1979). Chunking in recall of symbolic drawings. *Memory & Cognition, 7*(2), 149–158. http://dx.doi.org/10.3758/bf03197595.
eTouchSciences. (2014a). About us, from http://www.etouchsciences.com/ets/t/aboutus.

eTouchSciences. (2014b). Get in touch with science and math, from http://www.etouchsciences.com/ets/.
Gee, J. P. (2008). Videogames and embodiment. *Games and Culture, 3*(3-4), 253–263.
Gee, J. P. (2009). Deep learning properties of good digital games: How far can they go? In U. Ritterfeld, M. J. Cody, & P. Vorderer (Eds.), *Serious games: Mechanisms and effects* (pp. 65–85). New York: Routledge.
Gentner, D. (1983). Structure mapping: A theoretical framework for analogy. *Cognitive Science, 7*, 155–170.
Gentner, D., & Holyoak, K. J. (1997). Reasoning and learning by analogy: Introduction. *American Psychologist, 52*(1), 32–34.
Gentner, D., & Markman, A. B. (1997). Structure mapping in analogy and similarity. *American Psychologist, 52*(1), 45–56.
Goldstone, R. L., Landy, D. H., & Son, J. Y. (2010). The education of perception. *Topics in Cognitive Science, 2*, 265–284.
Hatano, G., & Inagaki, K. (1986). Two courses of expertise. In H. Stevenson, H. Azuma, & K. Hakuta (Eds.), *Child development and education in Japan* (pp. 262–272). New York: W. H. Freeman and Company.
Hektner, J. M., Schmidt, J. A., & Csikszentmihalyi, M. (2007). *Experience sampling method: Measuring the quality of everyday life*. Thousand Oaks, CA: Sage.
Hidi, S., & Renninger, K. A. (2006). The four-phase model of interest development. *Educational Psychologist, 41*(2), 111–127.
Hollender, N., Hofmann, C., Deneke, M., & Schmitz, B. (2010). Integrating cognitive load theory and concepts of human-computer interaction. *Computers in Human Behavior, 26*, 1278–1288. http://dx.doi.org/10.1016/j.chb.2010.05.031.
Holyoak, K. J. (1985). The pragmatics of analogical transfer. In G. H. Bower (Ed.), *The psychology of learning and motivation* (vol. 19, pp. 59–87). New York: Academic Press.
Holyoak, K. J. (2012). Analogy and relational reasoning. In K. J. Holyoak & R. G. Morrison (Eds.), *The Oxford handbook of thinking and reasoning* (pp. 234–259). New York: Oxford University Press.
Holyoak, K. J., Gentner, D., & Kokinov, B. N. (2001). Introduction: The place of analogy in cognition. In D. Gentner, K. J. Holyoak, & B. N. Kokinov (Eds.), *The analogical mind: Perspectives from cognitive science* (pp. 1–20). Cambridge, MA: MIT Press.
Holyoak, K. J., & Thagard, P. (1989). Analogical mapping with constraint satisfaction. *Cognitive Science, 13*, 295–355.
Holyoak, K. J., & Thagard, P. (1997). The analogical mind. *American Psychologist, 52*(1), 35–44.
Hummel, J. E., & Holyoak, K. J. (1997). Distributed representations of structure: A theory of analogical access and mapping. *Psychological Review, 104*(3), 427–466.
Jonassen, D. H. (2006). On the role of concepts in learning and instructional design. *Educational Technology Research & Development, 54*(2), 177–196.
Koedinger, K. R., Corbett, A. T., & Perfetti, C. (2012). The knowledge-learning-instruction framework: Bridging the science-practice chasm to enhance robust student learning. *Cognitive Science, 36*(5), 757–798. http://dx.doi.org/10.1111/j.1551-6709.2012.01245.x.
Lakoff, G. (1987). *Women, fire, and dangerous things: What categories reveal about the mind*. Chicago: The University of Chicago Press.
Lederman, S. J., & Klatzky, R. L. (1997). Relative availability of surface and object properties during early haptic processing. *Journal of Experimental Psychology: Human Perception and Performance, 23*, 1680–1707.
Linn, M. C., Davis, E. A., & Bell, P. (Eds.), (2004). *Internet environments for science education*. Mahwah, NJ: Lawrence Erlbaum Associates.
Massimini, F., & Carli, M. (1988). The systematic assessment of flow in daily life. In M. Csikszentmihalyi & I. S. Csikszentmihalyi (Eds.), *Optimal experience: Psychological studies of flow in consciousness* (pp. 266–287). New York: Cambridge University Press.

Merrill, M. D. (2002). First principles of instruction. *Educational Technology Research & Development, 50*(3), 43–59.
Miller, G. A. (1956). The magical number seven, plus or minus two: Some limits on our capacity for processing information. *Psychological Review, 63*, 81–97.
Mislevy, R. J., Behrens, J. T., Dicerbo, K. E., Frezzo, D. C., & West, P. (2012). Three things game designers needs to know about assessment. In D. Ifenthaler, D. Eseryel, & X. Ge (Eds.), *Assessment in game-based learning: Foundations, innovations, and perspectives* (pp. 59–81). New York: Springer.
Moneta, G. B. (2012). On the measurement and conceptualization of flow. In S. Engeser (Ed.), *Advances in flow research* (pp. 23–50). New York: Springer Science+Business Media.
Moore, D. R., Williams, R. L., II, Luo, T., & Karadogan, E. (2012). Elusive achievement effects of haptic feedback. *Journal of Interactive Learning Research, 24*(3), 329–347.
National Research Council (2011). Learning science through computer games and simulations. Committee on Science Learning: Computer Games, Simulations, and Education. In M. A. Honey & M. L. Hilton (Eds.), *Board on science education, division of behavioral and social sciences and education*. Washington, DC: National Academies Press. Retrieved from the National Academies Press website: *http://www.nap.edu/catalog.php?record_id=13078.*
Newell, A. (1990). *Unified theories of cognition*. Cambridge, MA: Harvard University Press.
Newell, A., & Simon, H. A. (1972). *Human problem solving*. Englewood Cliffs, NJ: Prentice-Hall, Inc.
Overliet, K. E., Krampe, R. T., & Wagemans, J. (2012). Perceptual grouping in haptic search: The influence of proximity, similarity and good continuation. *Journal of Experimental Psychology: Human Perception and Performance, 38*(4), 817–821.
Paas, F. G. W. C., Renkl, A., & Sweller, J. (2003). Cognitive load theory and instructional design: Recent developments. *Educational Psychologist, 38*(1), 1–4.
Paas, F. G. W. C., & Sweller, J. (2012). An Evolutionary upgrade of cognitive load theory: Using the human motor system and collaboration to support the learning of complex cognitive tasks. *Educational Psychology Review, 24*(1), 27–45. http://dx.doi.org/10.1007/s10648-011-9179-2.
Park, J., Kim, K., Tan, H. Z., Reifenberger, R., Bertoline, G., Hoberman, T., et al. (2010). An initial study of visuohaptic simulation of point-charge interactions. In *Paper presented at the IEEE haptics symposium, Waltham, MA*. March 25-26.
Polya, G. (1954). *Mathematics and plausible reasoning: Volume 1: Induction and analogy in mathematics: (Vol. 1)*. Princeton, NJ: Princeton University Press.
Reese, D. D. (2007). First steps and beyond: Serious games as preparation for future learning. *Journal of Educational Media and Hypermedia, 16*(3), 283–300.
Reese, D. D. (2009). Structure mapping theory as a formalism for instructional game design and assessment. In B. Kokinov, K. Holyoak, & D. Gentner (Eds.), *New frontiers in analogy research: Proceedings of the 2nd international conference on analogy (Analogy '09)* (pp. 394–403). Sofia, Bulgaria: New Bulgarian University Press. Available at: http://www.nbu.bg/cogs/analogy09/proceedings/42-T2.pdf.
Reese, D. D. (2010). Introducing Flowometer: A CyGaMEs assessment suite tool. In R. V. Eck (Ed.), *Gaming & cognition: Theories and perspectives from the learning sciences* (pp. 227–254). Hershey, PA: IGI Global.
Reese, D. D. (2012). CyGaMEs: A full-service instructional design model harnessing game-based technologies for learning and assessment. In L. Moller & J. B. Huett (Eds.), *The next generation of distance education: Unconstrained learning* (pp. 157–170). New York: Springer.
Reese, D. D. (2014). Digital knowledge maps: The foundation for learning analytics through instructional games. In D. Ifenthaler & R. Hanewald (Eds.), *Digital knowledge maps in education: Technology-enhanced support for teachers and learners* (pp. 299–327). New York: Springer.
Reese, D. D. (2015a). Affect during instructional video game learning. In S. Tettegah, & W. D. Huang (Eds.), *Emotions, technology, and games* (pp. 231–287). New York: Elsevier.

Reese, D. D. (2015b). Embodied learning systems. In J. M. Spector, T. Johnson, D. Ifenthaler, W. Savenye, & M. Wang (Eds.), *SAGE encyclopedia of educational technology*. Thousand Oaks, CA: Sage.

Reese, D. D., Seward, R. J., Tabachnick, B. G., Hitt, B., Harrison, A., & McFarland, L. (2012). Timed Report measures learning: Game-based embedded assessment. In D. Ifenthaler, D. Eseryel, & X. Ge (Eds.), *Assessment in game-based learning: Foundations, innovations, and perspectives* (pp. 145–172). New York: Springer.

Reese, D. D., Tabachnick, B. G., & Kosko, R. E. (2013). Video game learning dynamics: Actionable measures of multidimensional learning trajectories. *British Journal of Educational Technology*, Advance online publication, http://onlinelibrary.wiley.com/doi/10.1111/bjet.12128/pdf. http://dx.doi.org/10.1111/bjet.12128.

Renninger, K. A., & Hidi, S. (2011). Revisiting the conceptualization, measurement, and generation of interest. *Educational Psychologist*, *46*(3), 168–184. http://dx.doi.org/10.1080/00461520.2011.587723.

Rose, J. (2012). *eTouchSciences*. Retrieved November 15, 2014 from, http://www.youtube.com/watch?v=clRbxsZhS7Y&feature=youtu.be.

Samman, S. N., & Stanney, K. M. (2006). Multimodal interaction. W. Karowski (Ed.), *International encyclopedia of ergonomics and human factors* (vol. 2, 2nd Ed.). Boca Raton, Florida: Taylor and Francis.

Schertenleib, A., & Candey, R. (2013). *Sonification*. Retrieved November 14, 2014 from, http://spdf.sci.gsfc.nasa.gov/research/sonification/sonification.html.

Selye, H. (1974). *Stress without distress*. New York: Lippencott.

Selye, H. (1975). Confusion and controversy in the stress field. *Journal of Human Stress*, *1*(2), 37–44.

Selye, H. (2013). The nature of stress. Best of Basal Facts: 1976-1987, 629-639. Retrieved from http://www.ishafiles.com/eBooks/TheNatureofStress.pdf (Reprinted from The nature of stress, Basal Facts, 7(1), 3-11, by Hans Selye, 1985).

Shams, L., & Seitz, A. R. (2008). Benefits of multisensory learning. *Trends in Cognitive Sciences*, *12*(11), 411–417. http://dx.doi.org/10.1016/j.tics.2008.07.006.

Shute, V. J. (2011). Stealth assessment in computer-based games to support learning. In S. Tobias & J. D. Fletcher (Eds.), *Computer games and instruction* (pp. 503–524). Charlotte, NC: Information Age Publishing.

Shute, V. J., & Ventura, M. (2013). *Stealth assessment: Measuring and supporting learning in games*. Cambridge, MA: MIT Press. Retrieved from the The John D. and Catherine T. MacArthur Foundation Reports on Digital Media and Learning website, http://mitpress.mit.edu/sites/default/files/titles/free_download/9780262518819_Stealth_Assessment.pdf.

Shute, V. J., Ventura, M., Bauer, M., & Zapata-Rivera, D. (2009). Melding the power of serious games and embedded assessment to monitor and foster learning: Flow and grow. In U. Ritterfeld, M. J. Cody, & P. Vorderer (Eds.), *Serious games: Mechanisms and effects* (pp. 295–321). New York: Routledge.

Spellman, B. A., & Holyoak, K. J. (1996). Pragmatics in analogical mapping. *Cognitive Psychology*, *31*(3), 307–346.

Sweller, J., Kirschner, P. A., & Clark, R. E. (2007). Why minimally guided teaching techniques do not work: A reply to commentaries. *Educational Psychologist*, *42*(2), 115–121.

Sweller, J., van Merriënboer, J. J. G., & Paas, F. G. W. C. (1998). Cognitive architecture and instructional design. *Educational Psychology Review*, *10*(3), 251–296.

TechConnectWV (2014). *WV success story: eTouchSciences*. Retrieved from: http://techconnectwv.org/wv-success-story-etouchsciences/, November 15.

van Gerven, P. W. M., Paas, F., van Merriënboer, J. J. G., & Schmidt, H. G. (2006). Modality and variability as factors in training the elderly. *Applied Cognitive Psychology*, *20*(3), 311–320.

Wallace, D. S., West, S. W. C., Ware, A., & Dansereau, D. F. (2010). The effect of knowledge maps that incorporate gestalt principles on learning. *The Journal of Experimental Education, 67*(1), 5–16.

Wesslein, A.-K., Spence, C., & Frings, C. (2014). Vision affects tactile target and distractor processing even when space is task-irrelevant. *Frontiers in Psychology, 5*(84), 1–13.

Wolfe, J. M., & Horowitz, T. S. (2004). What attributes guide the deployment of visual attention and how do they do it? *Nature Reviews Neuroscience, 5*(6), 495–501.

Wolfe, J. M., Kluender, K. R., Levi, D. M., Bartoshuk, L. M., Herz, R. S., Klatzky, R. L., et al. (2011). *Sensation and perception* (3rd ed.). Sunderland, MA: Sinauer Associate, Inc.

CHAPTER 7

Designing Interaction Strategies for Companions Interacting with Children

Néna Roa Seïler
School of Computing, Center for Interaction Design, Edinburgh Napier University, Edinburgh, UK

INTRODUCTION

Companions are the evolution of Embodied Conversational Agents (ECAs) that are endowed with emotional abilities rendering them capable of establishing social and affective relationships with humans, to care for and provide companionship for them. The latter has also emerged and is recognized as a key factor in children acquiring and developing their social skills. The concept of such Companions is constrained by technologies that are emergent and cannot cope with the complexities and practicalities needed for the design and implementation of the detection and generation of affective behaviors within Companions. Difficult to use in real time, the impact of Companions' interaction with and on users is unknown or assessing it is problematic.

In this study, the challenge is to design and implement a set of Interaction Strategies for Companions acting as team-players in a Collaborative Collocated Learning Game with a group of children. Recent studies of Companions interacting with children direct the focus toward the embodiment, the materiality of these entities, how this affected the unfolding of their relationship, and toward the emotional response these emotionally intelligent, aware Companions elicited in the course of interacting with them (Roa-Seïler et al., 2014; Heylen et al., 2011).

We present some state-of-the-art Companions and describe how the Wizard of Oz (WoZ) protocol can be used to develop and test Conversational, Empathetic, and Domain Specific Interaction Strategies for them.

AGENTS AS COMPANIONS

A Companion is defined as a "robot or a virtual conversational agent that possesses a certain level of intelligence and autonomy, as well as social skills that allow it to establish and maintain long-term relationships with users" (Lim, 2012, p. 242). The term is meant to "invoke personification and anthropomorphism, and encompassing the widest possible range of devices and forms of interaction that, woven together, produce a relationship-building experience for people" (Benyon & Mival, 2013, p. 72). The idea of agents as Companions, machines with identifiable personalities was first proposed by Wilks (2005), as "intelligent, personalized, persistent, multi-modal interfaces for the Internet" (Wilks, 2006).

This initial "concept Companion"—an intelligent, emotionally aware entity—was driven by user expectations and facilitated by rapid progress in digital technologies. The expectations are that Companions sustain conversation over a reasonable length of time and touch on every aspect of their owner's daily life. They are to live in close proximity with humans, care for their well-being, serve their interests, do work on their behalf, and acquire extensive personal knowledge through interactions with them or from public data (Wilks, 2010). Their primary aim is to provide companionship.

COMPANIONS, CHILDREN AND COMPANIONSHIP

Companionship is part of the hierarchy of social needs along with love and belonging (Maslow, 1943). It implies a long-term relationship with a degree of empathy and intimacy and is particularly important in promoting children's healthy emotional and social development. Children not coping well with friendship and companionship issues will likely suffer loneliness, victimization by peers, poor adjusting to school, and/or engage in deviant behaviors (Elias, 1997; Ten Dam & Volman, 2007).

The value of human-animal companionship, that special sense of belonging and mutual need that exists in pet-human relationships is already proven as an aid to coping better with daily living or when facing difficult life stressors (McConnell, Brown, Shoda, Stayton, & Martin, 2011). The notion of an artificial Companion as a pet is not a farfetched one and would be readily accepted. Children abound in the capacity to transform the mundane into fantastic narratives, to attribute personality and life to inanimate toys and readily conjure up imaginary companions, invisible friends, animal or human, which to them are real and interact with them as such. In addition to being a source of enjoyment, entertainment, and of pleasurable moments, imaginary companions are a great support, good company, there as a friend, playmate, someone to talk to any time, any place, an aid to overcoming boredom and loneliness, and last but not least, to help children escape reality (Majors, 2013).

CURRENT COMPANION PROJECTS

The potentiality of these artifacts in a variety of roles accommodating children's needs has been demonstrated by several interesting prototypes, mostly as research projects. Many European funded research projects are devoted to that: the LIREC Project[1] explores how we live with digital and interactive Companions. More recent projects are learning-related to fulfill European Union objectives. The EMOTE Project[2] is a technology-enhanced learning system acting in the role of an empathy-based robotic tutor—"exploring how the exchange of emotional cues with an artificial tutor can create a sense of connection and act as a facilitator of the learning experience" (Emote project, 2014). Leite et al. (2010) studied the impact of synthetic expressions of empathy of an iCat robot in the role of an empathetic chess player and it was perceived as more friendly than an ordinary iCat robot, confirming that empathy plays a key role in an interaction.

The Huggable is a robotic research platform for pediatric care with the embodiment of an anthropomorphic Teddy Bear fantasy animal, skilled at relational and affective touch-based interaction with a person (Stiehl et al., 2009; Lee, Stiehl, Toscano, & Breazeal, 2009).

Adam, Cavedon, and Padgham (2010) propose a children Companion model based on the enjoyability of interaction, focusing on three features that interactive systems must have in order to engage the user. First, users should feel in control by customized and appropriate feedback. Second, the demands on users should be adjustable to their skills, that interaction should be challenging, but not overpowering. Third, the system should support social interaction. Empathetic personality and social skills are particularly relevant if they are to help, promote the owner's learning, and to engage with, establish, and maintain a long-term relationship with him.

Ideally, to nurture a collaborative relationship, establish a bond between user/player and helper, agreeing goals and the tasks to be undertaken to achieve them, adopting a nondirective attitude, and leaving the direction of control with the "client" (Raskin, Rogers, & Witty, 2007; The British Association for person-centred approach, 2013).

---

[1] http://lirec.eu/project
[2] http://www.emote-project.eu/

THE AFFECTIVE CHANNEL: A FRAMEWORK FOR EMOTIONALLY INTELLIGENT COMPANION INTERACTION

To have a Companion capable of the kind of engaging relationship discussed above, it needs a minimum set of tools to deliver the specific elements of interaction—to maintain conversation for a reasonable length of time, in this case with children, recognize and interpret their emotional states, and to respond to them empathetically using appropriate voice qualities, facial expressions, gaze, body language, and semantics (Roa-Seïler, Benyon, & Leplâtre, 2009; Roa Seïler, 2015; Smith et al., 2011).

Humans use these powerful communicative elements, intentionally or otherwise, to express, share, and transmit emotional information in face-to-face conversation, and the "Affective Channel" (AC) is the conceptual tool, the "affective intelligence toolbox" that endows Companions with "emotive aptitude" to emulate the human process and create the desired empathetic, engaging interaction. The AC integrates three phases—in Phase 1 the entire User Emotional State (UES) is identified and interpreted via Multimodal Sensory Detection systems. UES input would normally comprise speech, its measurable elements, prosody, tone of voice, pitch, speed of delivery, and semantic components and gestures, facial expressions, head position, gaze, body position (torso and hands), and touch (1, 2, 3 in Figure below (Figures 7.1 and 7.2)).

Phase 2, the Companion Interface (4), has at its heart the Emotion Interface Process (EIP) (5) that extracts the information, the UES, from the various input modalities, sorts, consolidates, and evaluates these (Figure 7.3). In Phase 3, the AC (6) using Cognitive Emotional Modeling and access to a global knowledge base converts these data into possible response elements or interaction strategies—constructs, attitudes, impressions—to match the UES, and is fed back to the Companion Interface for transmission to the user, completing the loop. For the system to recreate human-human interaction faithfully in real time and respond appropriately, all the stages, from detection to assessment, to response formulating, demand accuracy (Figure 7.4).

DESIGNING INTERACTION STRATEGIES FOR COMPANIONS

The AC output was organized as Emotional, Conversational, and Domain Specific, the latter to match their owner's needs profile. In this paper, the slant is toward children learning and playing, but the strategies can be applied in any setting with adults as well. The starting point is that face-to-face conversation

Designing interaction strategies 133

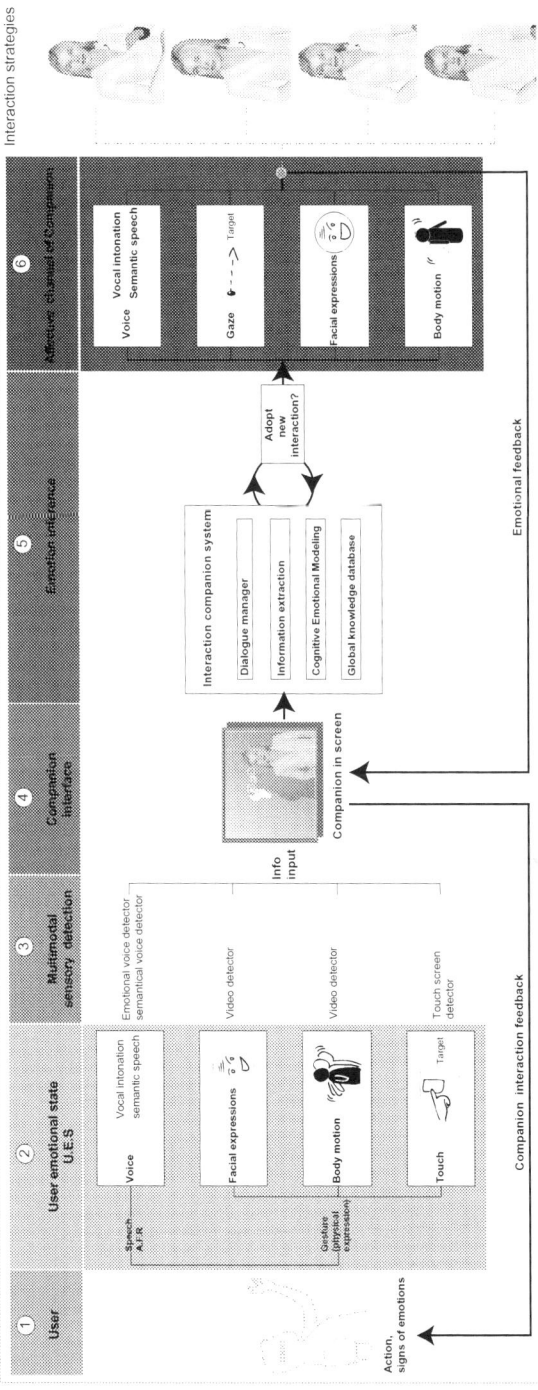

Figure 7.1 The overall process when interacting with an ECA. Courtesy of Dr. Néna Roa Seiler, unpublished doctoral dissertation.

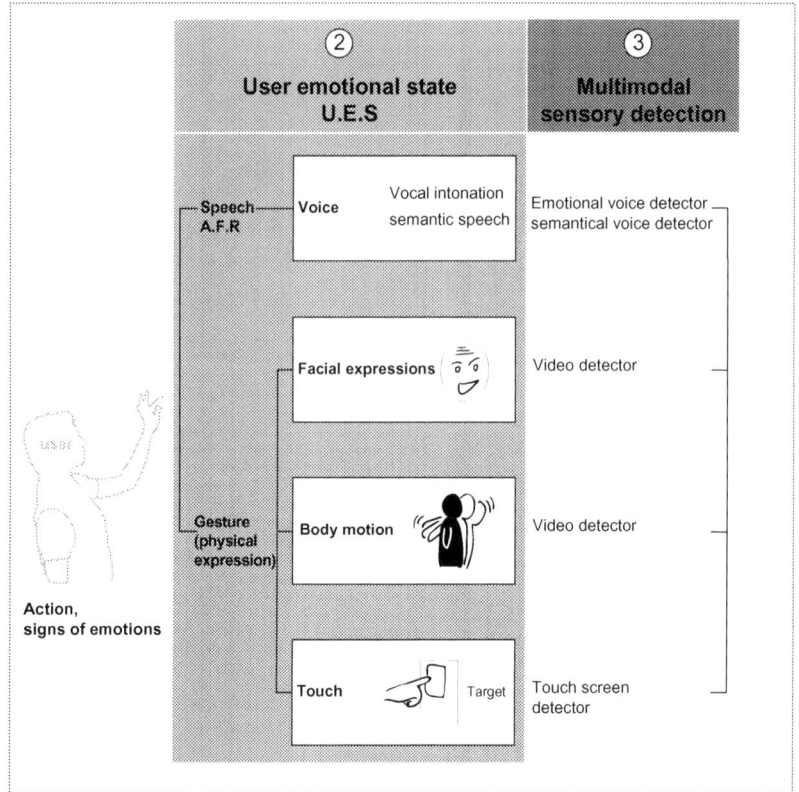

Figure 7.2 Phase 1 of the Affective Channel model. *Courtesy of Dr. Néna Roa Seiler, unpublished doctoral dissertation.*

is the human interaction strategy with effective communication as its end product; there is a continuous process of mutual listening, understanding, processing, and responding appropriately. Facial expressions are an integral part of communication, a nonverbal modality that enriches communication by complementing and enhancing the verbal content being conveyed (Schröder, 2006; Thiebaux, Lance, & Marsella, 2009).

The challenge is to develop like strategies for Companions, ones that would cope with face-to-face oral communication in the same manner, rather like people learning a foreign language apply and rely on some sort of strategy to convey or understand a message successfully, so conversation can progress (Somsai & Intaraprasert, 2011).

Previous agents-related research has been looked at to help our modeling of Companions. Cassel, Sullivan, Prevost, and Churchill (2000), when implementing the virtual Real Estate Agent, REA's behavior, investigated

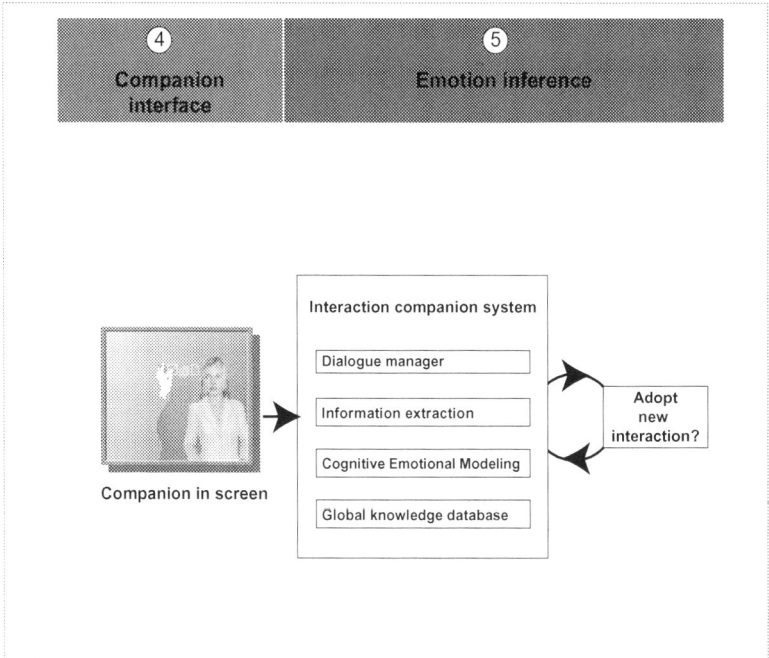

Figure 7.3 Phase 2 of the Affective Channel model. *Courtesy of Dr. Néna Roa Seïler, unpublished doctoral dissertation.*

the difference between the conversational function and its behavioral realization by studying the polysemous value of some multimodal cues based on the social rules of conversation such as turn taking, feedback with appropriate use of gaze, and facial backchannels. Backchannels, important elements of verbal and nonverbal behavior executed by the speaker (Maatman, Gratch, & Marsella, 2005), are gestures or sounds by the listener to give impetus and continuity to the conversation, an indication of their engagement or nonengagement in the interaction.

Greta was another interesting two-way, listener and speaker model of Conversation Interaction Strategy (De Sevin et al., 2010). Another is the "How Was Your Day" (HWYD) Companion prototype with its Short Loop and Main Loop Interaction Strategies facilitating the implementation of complete end-to-end affective conversation. These processes rely on information extraction by a Natural Language User (NLU) model with a Dialogue Manager, on an Emotional Model to formulate the user's mood in the Affective Strategy module (Benyon, Gamback, Hansen, Mival, & Webb, 2013; Smith et al., 2011).

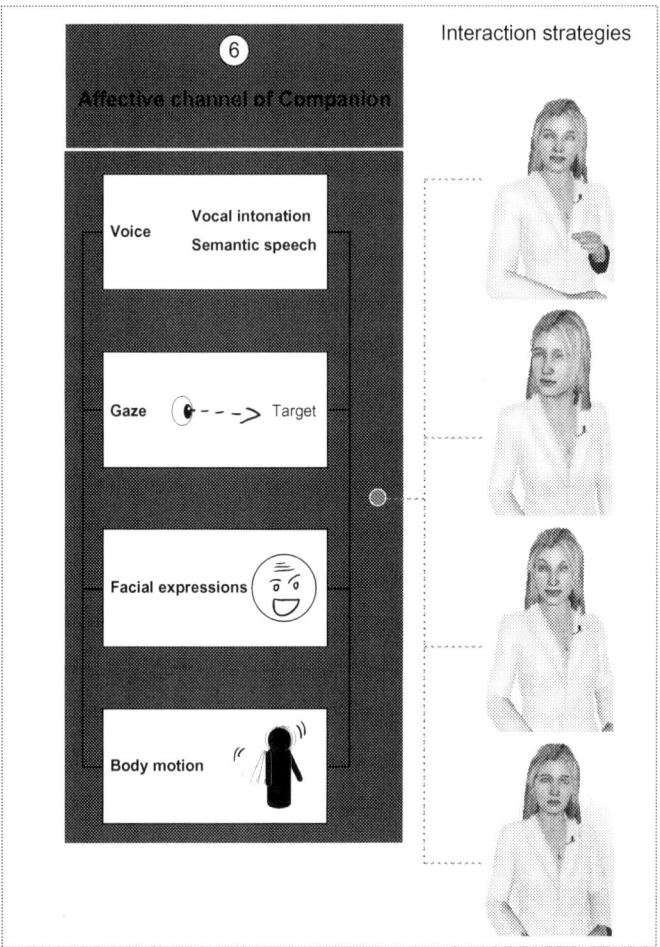

Figure 7.4 Phase 3 of affective channel model. *Courtesy of Dr. Néna Roa Seïler, unpublished doctoral dissertation.*

For the purposes of this work, the key parts of conversational behavior had to be in place, verbal and nonverbal signals accompanying and complementing each other in the exchange, commencing or ending a dialog, taking turns, handling interruptions, misunderstandings, feedback, that is, the modifiers, regulators of interaction that elevate it to a level where the exchange of multiple levels of information in real time is taken for granted. These strategies will be integrated in a WoZ simulation system to observe interaction in real time.

EMOTIONAL INTERACTION STRATEGIES

These strategies, below, are in line with users' requests for empathy in Companions—the key to emotional connection between beings (Roa Seïler, 2015; Rogers, 1995)—for it assists and supports the collaborative alliance (Bickmore & Picard, 2005) and instrumental in creating and maintaining long-term relationships (Raskin et al., 2007).

1. Sympathetic Strategy—is to offer an array of affective support—care, compassion, concern, consideration, kindness, and understanding—to combat the user's negative emotional state, modify it, and guide it toward a neutral or a positive one.
2. Cheerful Strategy—is one to encourage the maintenance the user's positive emotional state—contentment, happiness—when this is detected, by celebrating and/or sharing it.
3. Inquisitive Strategy—is a subcategory to the above two for determining the cause of the user's, player's negative or positive emotional state when it is not clear from the semantic analysis. This same strategy can be used when there is a doubt in the input detection process of the AC.

CONVERSATIONAL INTERACTION STRATEGIES (CIS)

The phases of systems' and humans' dialog are well defined as start and end of dialog, turn management (take turns to listen and take turns to speak), active waiting, and interruptions (Lopez Mencia, 2011). CIS are designed for resolving failures in the dialog system—not understanding, clarifying information, eliminating incongruences related to the user model (misunderstanding)—and for dealing with problematic conversational features such as listening after ceding a turn or being polite when interrupted.

1. Surprised Strategy—is to respond to input that is understood but not expected; used when the information provided by user doesn't match the global knowledge database of the Companion's User Model in the EIP (see above) or when the user's choice contradicts the User Model, the Companion's knowledge of the user.

2. Confused Strategy—is to respond to input that is not understood, the Dialogue System has crashed and the information is garbled or when the user gives contradictory or incomplete information. The user may be describing somebody as—"He has a beard"—and the system doesn't have enough information to recognize whether it is a "beard" or a "bird."
3. Listening Strategy—is to make the most of turn management; the user is saying something, the Companion expresses interest by using only short backchannel noises or phrases, indicating that she is listening with attention thereby encouraging the user to continue speaking, all the while collecting information about the user. These could be interjections or short statements (e.g., Ouch! Shhh. Wow! Aha! Bravo! Great!).
4. Idle or Stand-by Strategy—is a strategy for looking interested and inviting, expecting and cajoling input from the silent user.
5. Interrupted Strategy—is a strategy to respond to the user interrupting. The Companion might stop talking and display a gesture to mean "Sorry," allowing the user to continue speaking. The full set of Interaction Strategies proposed for Companions is presented in Table 7.1.

DOMAIN SPECIFIC STRATEGY

These are interaction strategies within the specific context in which the Companion needs to be a helpful performer to its owner. They are responses to the user's environment or the context (model or script) of the interaction. There may be several steps depending on the specificity of the role to be performed, for example learning, playing, etc. In this work, the domain specific is Collaborative Learning as shown in the Table 7.2.

Having designed the AC, the interaction strategy resource of the Companion, it was then tested in a Collaborative Learning environment setting, implemented in a Serious Game, which is a computer application, an educational scenario run in a video game (Alvarez, 2007). Such a setting benefits participants as they learn, get particular information from one another, exchange resources and skills, and evaluate each other's proposals for problem solving (Chiu, 2000). That it is playing a game also reduces anxiety associated with the learning process and pupils learn unaware of making an effort (Mouaheb, Fahli, Moussetad, & Eljamali, 2012). The WoZ will be used to test a simplified AC model of emotional responses with the children (see below).

Table 7.1 Interaction strategies for companions

Interaction strategies		Description
Emotional	Sympathetic	Responding to a negative emotional state of a user as detected by his facial expression and/or by the semantic and/or emotional analysis of the user statement.
	Cheerful	Responding to a positive statement from the user as detected by his facial expression and/or by the semantic and/or emotional analysis of the user statement.
	Inquisitive	Subsequent state to "Sympathetic" or "Cheerful" which attempts to determine cause of the user's negative or positive emotional state.
Conversational	Surprised	Responding to input (verbal or Non verbal) from user that is understood but unexpected.
	Confused	Responding to input from user that is not understood.
	Listening	Listening to input from the user.
	Idle	Waiting for input from the user.
	Interrupted	Responding to the user interrupting the ECA's speech.
Domain specific	Responsive	Responding to user statement depending on interaction environment or context (model or script).

Courtesy of Dr. Néna Roa Seïler, unpublished doctoral dissertation.

Table 7.2 Collaborative learning interaction strategies for companions

Domain specific	Interaction strategy	Description
Collaborative learning interaction strategy	Facilitator	Invites participation. Monitors the group's progress. Encourage group's harmony (by tempering conflicts, building compromises, etc).
	Proposer	Suggests new ideas.
	Supporter	Evaluate proposal. Looking at advantages and disadvantages.
	Critic	Identifies weaknesses or errors. Suggested alternatives. Challenges the original claim (why this or that).
	Reminder	Summarizes the group's progress.

Courtesy of Dr. Néna Roa Seïler, unpublished doctoral dissertation.

WIZARD OF OZ EXPERIMENT

The WoZ is a research experiment in which people interact with a system they believe to be autonomous, but one that is actually being at least partially operated by an unseen human being (Bradley, Mival, & Benyon, 2009; Martínez García, Craig, Roa-Seiler, & Benítez Saucedo, 2012). It is a powerful technique for interacting in real time, hitherto not open to Companions due to lack of suitable technology. The WoZ is ideal when working with children as it is free of slow or inappropriate responses and other previously encountered undesirable aspects. A modified AC was implemented with only Facial and Vocal expressions in the Emotional and Conversational Interaction Strategies.

Facial expressions are mostly automatic, powerful indicators of human emotions, helpful in the building of relationships with others (Keltner, Ekman, Gonzaga, & Beer, 2003; Lakin and Chartrand, 2003). When creating a visual database for each interaction strategy presented in Table 7.1, it was decided to work with trained actors adept at reproducing natural and accurate representation of emotions. Following Busso and Narayanan (2008) guidelines, scripts, a series of small scenarios linked to an emotional or a conversational situation created to support the Companion's Emotional and Conversational strategies, were to be rehearsed. Facial expressions were qualitatively evaluated and rated in intensity using a Likert scale (Figure 7.5).

Focus groups were organized to suggest the best verbal responses, phrases, and interjections for the Interaction Strategies. The appropriateness of matched Facial expressions and Verbal statements data gathered for each Interaction Strategy was qualitatively evaluated using the Plutchick model of Emotions (Roa Seïler, 2015).

The system paired the Facial expressions consecutively with the series of Verbal statements (see Table 7.3) to arrive at matching "sets" of synthetic facial expressions with the same characteristics as spontaneous, natural human facial ones to be implemented in the Conversational and Emotional Interaction Strategies.

The Wizard system also integrates the Domain Specific Strategy involving children in a Collaborative Learning Game environment. A pilot session with Mexican teachers was organized to obtain some verbal strategy patterns. A cheerful facial expression displaying good intentions and willingness were used for all states of Collaborative Learning Interaction Strategy.

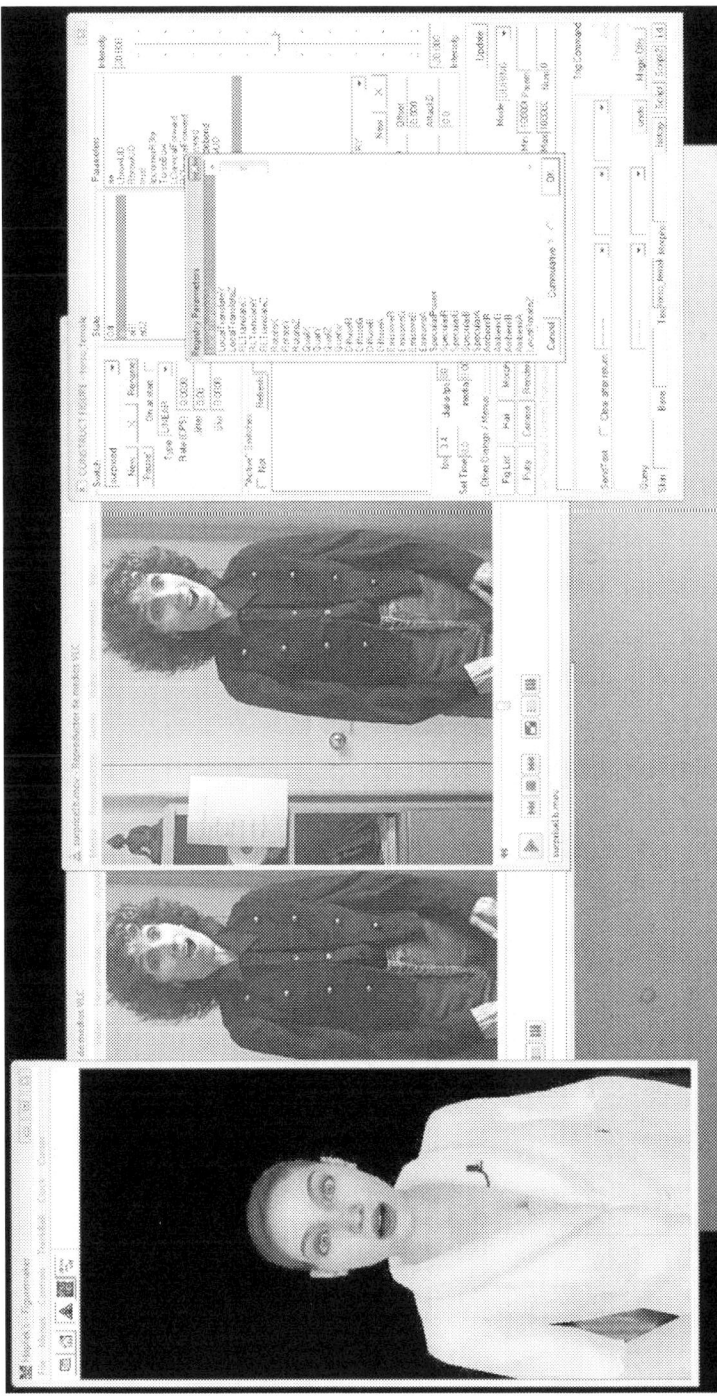

Figure 7.5 Process of making facial expressions for a virtual companion. *Courtesy of Dr. Néna Roa Seïler, unpublished doctoral dissertation.*

Table 7.3 Facial expressions and verbal statements for each interaction strategy

Interaction strategy	Facial expression	Statement
Sympathetic	9	44
Cheerful	8	78
Inquisitive	8	15
Surprised	15	12
Confused	9	4
Listening	7	12
Interrupted	8	6
Idle	1	

Courtesy of Dr. Néna Roa Seïler, unpublished doctoral dissertation.

THE EXPERIMENT

This experiment investigates how children perceive Companions in learning game environments, how they construct a social relationship with them, and their interaction preferences with them. This was run using a Video game platform for children in fourth grade, with a virtual Companion in three different embodiments. Children's perception of Companions in terms of physical appearance, personality, and functionality/utility was observed and noted and their feelings toward them was measured by prEmo, a nonverbal tool that measures emotions elicited by products. By this medium, instead of relying on words, children can represent their emotions using expressive cartoon animations (Desmet, 2005).

THE SUBJECTS OF THE EXPERIMENT

Twenty-four students, 9 to 10 years old from the rural school of Acatlima in Huajuapan de Leon, Oaxaca, Mexico took part in the experiment, 14 girls and 10 boys. Permission from the school principal and parents was obtained, the observational nature of the research and the concept of Companions explained to administrative personnel, parents and teachers, and finally the screen-based companion Samuela and the robot Nao were introduced. Samuela is a 3D screen-based character, Nao is a robot and the third *embodiment* was Ari, an actual person made to look like a cartoon character and projected on the screen as shown in Figure 7.6.

Samuela and Nao interacted with the children via the WoZ and Ari, the Human Research Assistant, through a screen-based cartoon image created by a filter placed on the camera viewing her face as the image above shows. The design of the experiment involved creating video games, setting up a WoZ system, planning the physical layout, and preparing the format of

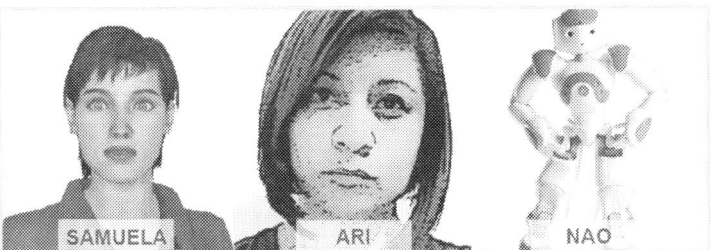

Figure 7.6 The three companions in the experiment. *Courtesy of Dr. Néna Roa Seïler, unpublished doctoral dissertation.*

Focus Groups and personal Interviews. This was a "within-subjects" experimental design, that is, each child participated in all the experiments with the three Companion embodiments.

THE GAME

The first task was to design a game, long recognized for its educational and motivational value—in this case a collaborative educational "Serious Game." A Serious Game is an educational scenario combining play and technology in a video game that can be applied in diverse areas from education, health, to scientific exploration, engineering, and many other domains (Alvarez, 2007). They are more than mere entertainment, the adding of pedagogical techniques to deliver educative content with players in control of their learning experience makes the experience less stressful and more productive. Education-related games are also called Edutainment.

Video games aren't senseless play spaces, as complex processes of learning, thinking, and social practices take place playing them (Barab et al., 2007). The player, engrossed in a fictional situation where emotions, imagination, fantasy, and thinking interplay, finds relevant learning metaphors or analogies. In fact, experts view video games as one of the "most significant forms of media for the enculturation of young people" (Barab, Thomas, Dodge, Carteaux, & Tuzun, 2005).

An example of a game with an educational intent with the additional objective of excluding violence is the Quest Atlantis Project (Barab et al., 2007), a 3D multiuser virtual environment where players explore different worlds, embark on missions of different levels of complexity, and face various challenges called Quests (atlantisremixed.org, 2014).

PlayPhysics is an emotional game-based learning environment developed at Trinity College Dublin for teaching Physics at undergraduate level, where the student is an astronaut charged with the mission to save his or her mentor who has been trapped on a space station. In the process, players learn, explore, and understand a whole range of physics-related concepts while suitable guidance is given to them according to their detected emotional state by the system (Muñoz, Mc Kevitt, Lunney, Noguez, & Neri, 2010).

To gather feedback for the design of the game, three brainstorming sessions were organized with two 4th and 5th grade school teachers, two multimedia Master's degree students, and two members of our research team. The teachers' suggestion was accepted that the game be used to support subject areas where children had the most difficulty, mathematics and reading and learning English, and finding out about the United States, as most of the children had one or both parents living there. In the end it was decided that there would be one multiplayer educational videogame designed to support mathematics learning.

The game's design priorities were that it would be nonviolent, stayed clear of gender stereotypes, and complied with cultural norms. Academically, it had to be readily accessible to all levels of learning as appropriate to age and present progressively increased degrees of difficulty. It was also considered suitable that elements of the children's Mixtec and Mexican cultural heritage with their vibrant and colorful hieroglyphics should be used so that children discover and identify with these, the powerful historic and symbolic inheritance in their life even today. These icons are no longer used in writing but still appear in logos, books, and textile designs (Craig, Roa-Seïler, Lara Rosano, & Martínez Díaz, 2013; Figure 7.7).

Figure 7.7 Images from the Mixtec Codices used as characters in the videogames: Jaguar, Eagle, Death and the Mixtec hero Ocho Venado. *Courtesy of Dr. Néna Roa Seïler, unpublished doctoral dissertation.*

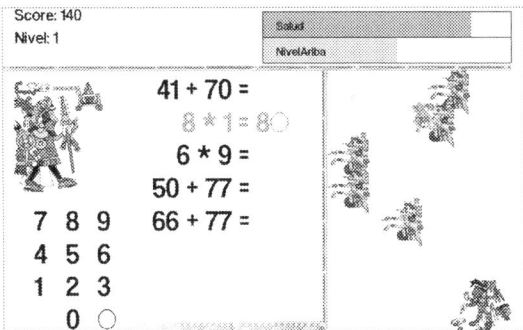

Figure 7.8 Screenshot of the Mathematics video game. *Courtesy of Dr. Néna Roa Seïler, unpublished doctoral dissertation.*

The Mathematics video game uses the metaphor of "tower defense"—students have to solve mathematical equations in order to fire eagles to prevent death figures from reaching the perimeter wall of their "tower" and drain their life-force. This game uses the strategy of defense rather than attack and it highlights the idea that good performance protects your wall. "Jaguar" represents the user on the left of the screen, in front is a list of sums. Below is a keypad for performing the operations. To the right of the sums is a vertical wall and beyond that are the death figures, which advance slowly from right to left. If one reaches the wall, it stops and begins to drain the life force of the user (Figure 7.8).

The life force bar is shown in red at the top and when it reduces to zero the game is over. To prevent this, the user solves sums picked using the keypad, and shoots eagles to stop the death figures touching the wall. Incorrect answers deplete the life force but when an eagle hits a death figure, it pushes it away from the wall and increases the user's score, raising the level of the life-force bar. When the bar level is completely green, every sum fires an eagle to push out all the death figures and the user progresses to the next level. As the level increases the speed of the approaching death figures increases too and the sums become gradually more difficult (Craig et al., 2013).

THE WIZARD OF OZ (WOZ) OF SAMUELA, NAO AND ARI

To run the experiments with the WoZ system, two separate rooms were set up: an Experiment Room and a Control Room (see Figure 7.9).

Three cameras installed in the Experiment Room followed the children's movements and facial expressions, recording their emotional states. Here, the pupils played the video games on the 42 inch multitouch screen, a camera

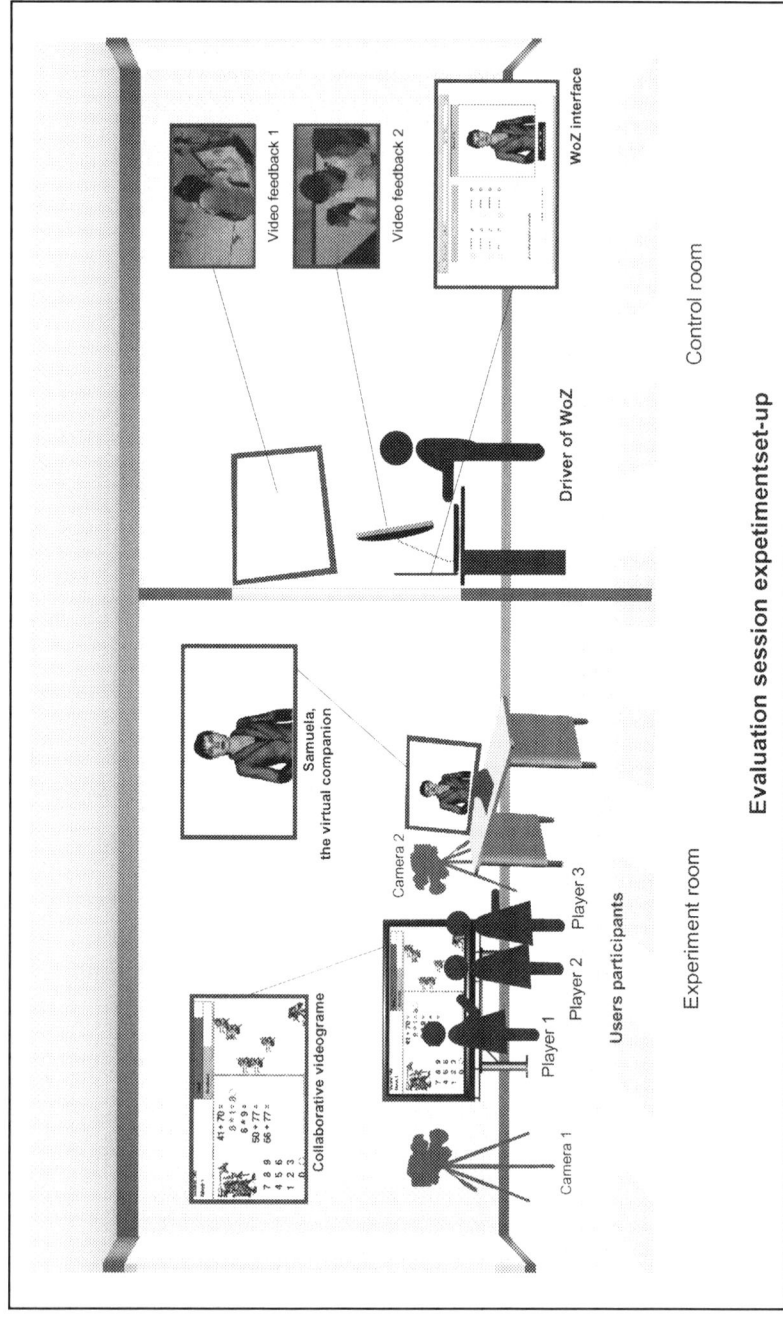

Figure 7.9 Evaluation session experiment set up. *Courtesy of Dr. Néna Roa Seiler, unpublished doctoral dissertation.*

installed on the screen keeping a record of the number of right answers. The rooms were separated by a two-way glass wall. All the Visual and Audio feedback from the Experiment Room was displayed in the Control Room allowing the Wizard located there to know what was happening. The Wizard was a Mexican Computer Science student manipulating the Companion's actions.

The WoZ interface comprised two windows, one showing the Control Board (CB) and the other Samuela's image. The Samuela window also appeared on a screen in the Experiment Room, the CB window remaining on the screen in the Control Room. Thus the Wizard could manipulate Samuela from the Control Room without the players knowing. The Conversational, Emotional, and Domain Interaction Strategies, as described earlier, were implemented.

At the top of the Wizard's CB are the three Emotional Interaction Strategies. The four Conversational Interaction Strategies are on the second row and on the right are the five roles of the Collaborative Learning Domain Interaction Strategy (Figure 7.10). The other buttons are "Welcome," "Goodbye," as well as a "Yes" and "No."

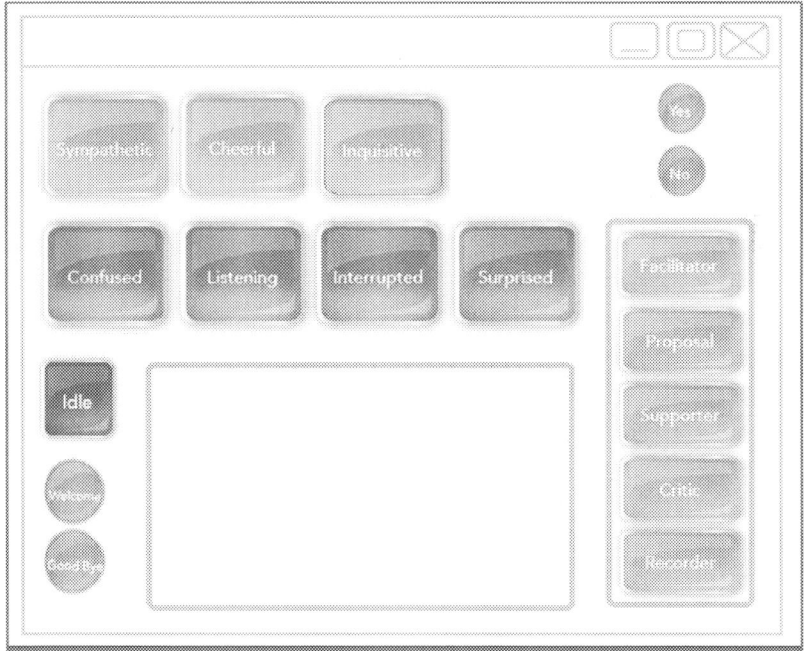

Figure 7.10 Screen shot of Wizard's control board of WoZ of collaborative co-located game. *Courtesy of Dr. Néna Roa Seïler, unpublished doctoral dissertation.*

At the bottom of the CB is a Text box in which the Wizard would write improvised phrases according to the interaction he was observing and the feedback he was getting from the cameras. For example, the Wizard would compliment a child if she played well, forcing the death figures to retreat. When the Wizard pushed the Cheerful Button, a Cheerful Statement would appear in the Text box. If the Wizard considered the statement inappropriate, he would push the button again until an appropriate phrase appeared which he would then personalize. So, if "How exciting!" is inappropriate, he would push the Cheerful Button again and the phrase "That's good!" would appear. He would then complete the phrase by writing "Well done, girls!," seeing that the team was comprised entirely of girls. The WoZ of the Nao robot companion was a very simple Text-to-Speech interface implemented in Phyton language. The human Ari Companion also interacted with the children in real time using human voice.

THE PILOT SESSION

A Pilot Session was set up with children from a different class, of the same age and grade and a teacher was invited to observe the session, the game and the way Samuela functioned. He offered several suggestions, one that Samuela remind the children often of the game tactics because that is how they work in class, one student charged with reminding the others what they are supposed to be doing. His other suggestions were to factor the larger numbers to make the mathematics problems easier and for Samuela to not give any answers because in so doing she would be acting as a teacher, this suggestion being particularly valuable as it confirmed the validity of the intention that a Companion should remain a Companion and not become a teacher.

The idea was to have a game geared to the children's needs with the Companion the medium of support in a learning environment, but not being an integral part of the game; the Companion does not act as a Learning Companion but as a Companion in a Learning Environment. The difference is crucial—a Learning Companion is a "Pedagogical Agent" (Arroyo, Woolf, Royer, & Tai, 2009), a task-oriented agent embedded in the game itself, whereas a "Companion in a learning environment" remains outside the game screen, part of a social relationship, companionship in particular, and capable of performing multiple tasks. It was rewarding to observe exactly how the children integrated the Companion into their teams and how its presence and its companionship influenced their learning performance.

Table 7.4 Dialog scripts and sentence patterns for collaboration roles inspired by Chiu (2000)
Collaborative learning domain
interaction strategies

Domain specific interaction strategies	Context of game or interaction
Facilitator	Hey, it's so and so's turn.
Proposer	I suggest you do the addition in three parts. For example, first 20 plus 30 plus 17.
Supporter	You said 5 times 40. Hmmm, that will be a pretty big number.
Critic	5 times 8... I don't think it's 35 ...
Reminder	Your life force bar is running low. Be careful!

Courtesy of Dr. Néna Roa Seïler, unpublished doctoral dissertation.

These dialog scripts and sentence patterns are only models, mere guidelines to refer to by the Wizard, because he could improvise drawing on his knowledge of each strategy and to adapt to what was happening in any given moment (Table 7.4).

THE FOCUS GROUPS AND INTERVIEWS

A series of focus groups and personal Interviews were organized with the children after they had played the game to solicit a list of adjectives that describe their perception of the three Companions, to rate them and then from the ratings derive a Semantic Differential (SD) for each. The personal Interviews, asking the children Free Recall questions (Markopoulos, Read, MacFarlane, & Hoysniemi, 2008), would help to refine the understanding of their perception. The focus groups would be conducted the day after the game session, the interviews the following day with assistance from four graduate students from the University, closer in age to the children to avoid their being intimidating by adult authority figures. Table 7.5 below displays the various steps in implementing the Experiment.

THE IMPLEMENTATION OF THE EXPERIMENT

The first day was dedicated to observing the children play the mathematics game. On the second, Adjectives to describe the children's perception of the Companions were gathered and rated and on the third the graduate students interviewed the children.

Table 7.5 Steps protocol of the experiment

Day	Step	Action	Time	Group
1	1	Arrival of children to the USALAB at Universidad Tecnológica de la Mixteca.	20 min	24
	2	Introducing the Companion to the children.	1-3 min	8(3)
	3	Playing the mathematics game with the Companion.	10 min	8(3)
2	4	Gathering Adjectives to qualify the Companion.	40 min	3(8)
3	5	Measuring the meaning of these Adjectives.	10 min	Individual
	6	Clarifying the meaning of these Adjectives and soliciting further information regarding Companion.	20 min	Individual written by the pollster.

Courtesy of Dr. Néna Roa Seïler, unpublished doctoral dissertation.

Step 1: Observing the Children Playing the Game

As the 24 children arrived at the Usability Lab at the Universidad Tecnológica de la Mixteca accompanied by four Master's degree students and two undergraduate students, they were welcomed into the boardroom, snacks and soft drinks provided, where they would watch movies when not involved in the game sessions.

Eight groups of three children were organized, their attitudes toward mathematics were recorded, the information passed to the Wizard who, knowing that a particular child didn't like mathematics would offer him support and special encouragement through the Companion.

The children knew nothing of the installations in the Control Room and in successive order each group of three children would be conducted to the Experiment Room to play the game. Those waiting to take their turn could not observe their classmates playing the game (Figure 7.11).

The experiment had three operational elements, a multiplayer educational video game connected to a multitouch screen, a Companion displayed by remote control, and the WoZ-operator who controlled the Companion.

One camera used a long shot, the other a closeup shot to follow the children's facial expressions, providing a permanent record of the game sessions. Users' correct responses and the number of right answers were also recorded. The set-up of the Collaborative collocated learning game is seen in Figure 7.12.

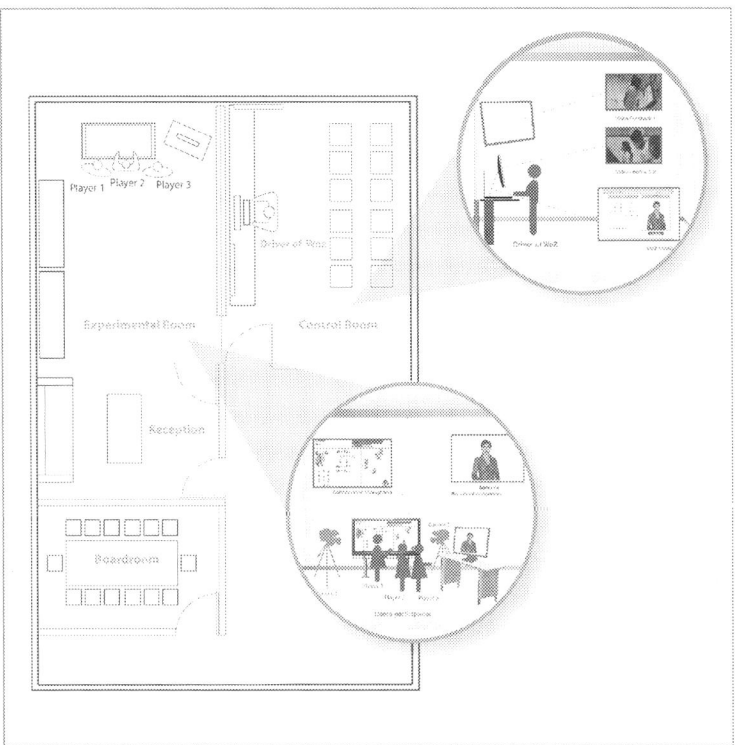

Figure 7.11 Layout of the experiment. *Courtesy of Dr. Néna Roa Seïler, unpublished doctoral dissertation.*

Each group of children played the game three times, each time with a different Companion following the same protocol. The Companions Samuela and Ari appeared on the right of the game screen, while the robot Nao was placed on a table on the same side of the screen.

The Companion invited the children to prepare a strategy to play the game (see Table 7.6) and to share roles to get a higher score. In order to avoid a sampling bias known as the "novelty effect"—the effect of stress, pleasure, or anticipation when confronted with something new for the first time—the pupils met each Companion in a randomized order.

Step 2: Determining the Children's Perception of the Companions via Focus Group Activities

The day after the game session pupils attended a focus group session led by one of the graduate students. The Guidelines for Research with Children

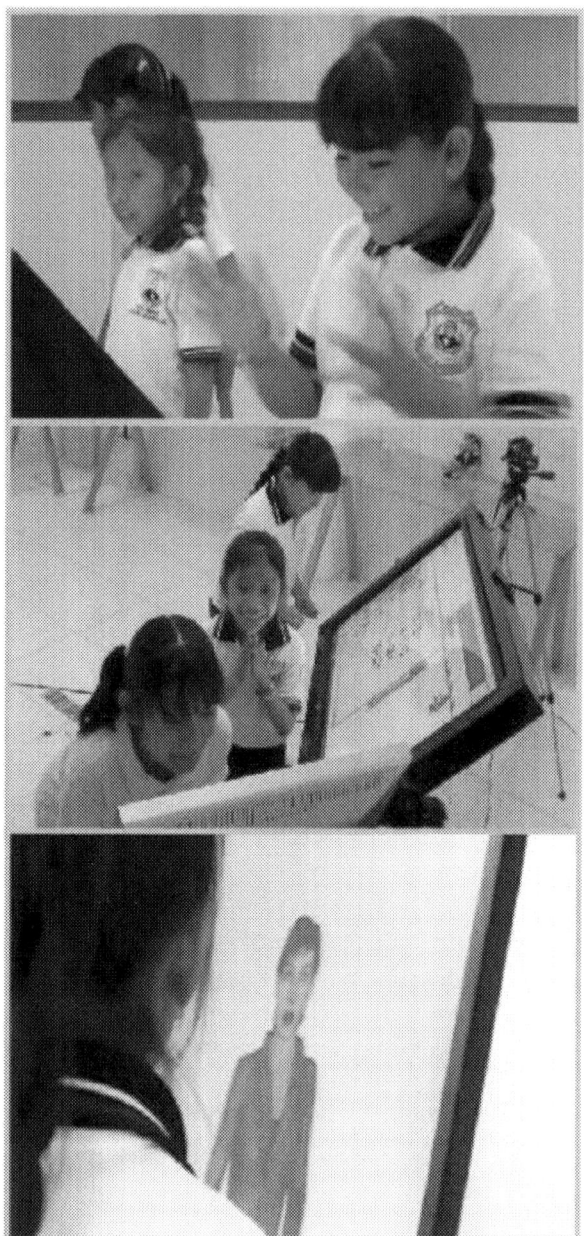

Figure 7.12 Children playing the collaborative co-located learning game. *Courtesy of Dr. Néna Roa Seïler, unpublished doctoral dissertation.*

Table 7.6 Introductory dialog scenario with Samuela

Speakers Who speak	Dialog	User emotional state	Interaction strategy
Samuela	Hi girls, how are you?		Greetings
1st Girl	Fine		
2nd Girl	Fine		
3rd Girl			
Samuela	Thank you for coming. What are your names?		Greetings Introduction
1st Girl	Angeles		
2nd Girl	Sarahi		
3rd Girl	Zulamita		
Samuela	I've been told that your mother has just had a baby.		
1st Girl	Yes, it's a girl. Her name is Hanna.	Positive emotion, smile.	Oh, I'm so happy to hear that. *Cheerful*
Samuela	And how are you girls?		Are you well?/ *Inquisitive*
2nd Girl	Yes	Positive emotion, smile.	
3rd Girl	Yes	Positive emotion, smile.	
Samuela	OK, my name is Samuela and I'm going to help you with the game. You can ask me any question you want, but it would be better if you prepared a strategy for how you are going to allocate and share the tasks; so that everyone gets a chance to play and we get the most life force possible. Do you see the life force there at the top of the screen? Every time you are correct it grows. When you make a mistake it gets smaller. So you should really concentrate so that it doesn't get smaller. When the life force bar reaches zero, the game is over. Do you understand?		

Continued

Table 7.6 Introductory dialog scenario with Samuela—cont'd

Speakers Who speak	Dialog	User emotional state	Interaction strategy
1st Girl	Yes	Positive emotion, smile.	
2nd Girl	Yes	Positive emotion, smile.	
3rd Girl	Yes	Positive emotion, smile.	
Samuela	OK, let's begin!!!		OK, let's begin!!! Proposer/ Domain

Courtesy of Dr. Néna Roa Seïler, unpublished doctoral dissertation.

and Young People (Shaw, Brady, & Davey, 2011) as to number of participants and to age range in a focus group were adhered to, with eight children, each session lasting for approximately 40 minutes. The graduate student showed the children a photo of each Companion first, asking them to provide three Adjectives to describe them in terms of physical appearance, personality, and functionality.

The most frequent Adjectives given were used to form a list of 22 "descriptors" paired with 22 opposite descriptors (e.g., pretty versus ugly, intelligent versus stupid, or human versus robotic). Table 7.7 shows the list of Adjectives and their opposites most applied by the children in Mexican Spanish with English translation. The children then scored each Companion according to these descriptors and the graduate students applied the SD to measure the connotative meaning of concepts providing information about their emotions, feelings, and attitudes that each word elicited (Hsu, Chuang, & Chang, 2000). The evaluation lasted 10 minutes and took place later the same day.

Step 3: Refining the Understanding of the Children's Perception via Interview Sessions

In the third phase of the experiment, the graduate students conducted 10 minute face to face interviews with the children using a 24-item Questionnaire with the objective of clarifying the meaning of the higher rated

Table 7.7 The global descriptors collected to qualify companions
Companions

Beautiful	⑦	⑥	⑤	④	③	②	①	Ugly
Thin	⑦	⑥	⑤	④	③	②	①	Fat
Fun	⑦	⑥	⑤	④	③	②	①	Boring
Pleasant	⑦	⑥	⑤	④	③	②	①	Unpleasant
Small	⑦	⑥	⑤	④	③	②	①	Large
Nice	⑦	⑥	⑤	④	③	②	①	Grumpy
Helpful	⑦	⑥	⑤	④	③	②	①	Unhelpful
Robotic	⑦	⑥	⑤	④	③	②	①	Human
Intelligent	⑦	⑥	⑤	④	③	②	①	Stupid
Loving	⑦	⑥	⑤	④	③	②	①	Unloving, cold
Good	⑦	⑥	⑤	④	③	②	①	Bad
Serious	⑦	⑥	⑤	④	③	②	①	Smiling
Human	⑦	⑥	⑤	④	③	②	①	Not Human
Real	⑦	⑥	⑤	④	③	②	①	Unreal
Old	⑦	⑥	⑤	④	③	②	①	Young
Technological	⑦	⑥	⑤	④	③	②	①	Non technological
Teaches	⑦	⑥	⑤	④	③	②	①	Doesn't teach
Beneficial	⑦	⑥	⑤	④	③	②	①	Harmful
Friend	⑦	⑥	⑤	④	③	②	①	Unfriendly
Teacher	⑦	⑥	⑤	④	③	②	①	Not Teacher
With teeth	⑦	⑥	⑤	④	③	②	①	Without teeth
Man	⑦	⑥	⑤	④	③	②	①	Woman

Courtesy of Dr. Néna Roa Seïler, unpublished doctoral dissertation.

descriptors, determine perceptions of appearance, trustworthiness, emotions, and empathy between the children and the Companions and to explore the children's familiarity with new technology. The Questionnaire questions are presented in Table 7.8.

Some of the questions were formulated as scenarios, generative questions, similar to those used in previous studies of children's perceptions (Vosniadou, 2002). These demand a precise answer to a new problem which can then be solved on the basis of previously saved information. To limit children writing when working (Markopoulos et al., 2008), the four graduate students conducted the face-to-face Interviews and completed the Questionnaires.

REPORT OF FINDINGS

The Questionnaire revealed that in spite of their economic situation, the children proved relatively tech-savvy. Twenty of the 24 children were familiar with the use of cell phones (83.3%), half of them had already used a tablet

Table 7.8 Questionnaire questions

Questionnaire

Inquiry	Description
1	5 questions to clarify Physical Descriptors of the Companions.
2	8 questions to clarify Personality Descriptors of the Companions.
3	3 questions to complement information about the functionality and helpfulness of the Companion.
4	3 question with scenarios about the kind of interaction children would like to have with the Companion.
5	1 question about the frequency of contact children would like to have with the Companion.
6	1 question about Companion's representation.
7	1 question about Companion' trustworthiness.
8	4 questions about emotions elicited by the game and by each Companion.
9	1 question with scenarios about Companion's empathy.
10	1 section with six questions to qualify children's knowledge about devices and new technology.

Courtesy of Dr. Néna Roa Seïler, unpublished doctoral dissertation.

(50%), 70.83% had used the desktop computer at school, and 75% played video games frequently. Only three students were not familiar with video games and only one was unfamiliar with cellphone technology. That is, children were not overwhelmed by the technologies used by Companions.

PERSONALITY DESCRIPTORS

In descending order of importance the personality descriptors were: friendly, good, real, funny, intelligent, loving, nice and pleasant, as shown in Figure 7.13 and examples are described below.

Friendly applies to someone who helps, listens, and pays attention to them—"Someone I can tell my deepest secrets to and he/she won't tell anyone." Ari was the friendliest Companion, likely based on her perceived human qualities, such as the emotional tone of her voice, qualities Nao and Samuela lacked. Good is the quality of a person who helps them, shares with them, and is nice to them —"A good person is one who cares for children," "one who is not mischievous." The three Companions were considered good because they helped solve operations. One child defined funny as a person who smiles, plays, and talks and Nao turned out to be the funniest tutor. Users gave nice to describe a person who is respectful and polite. Confusion arose around the word *simpatico* which was understood as sympathetic, but the

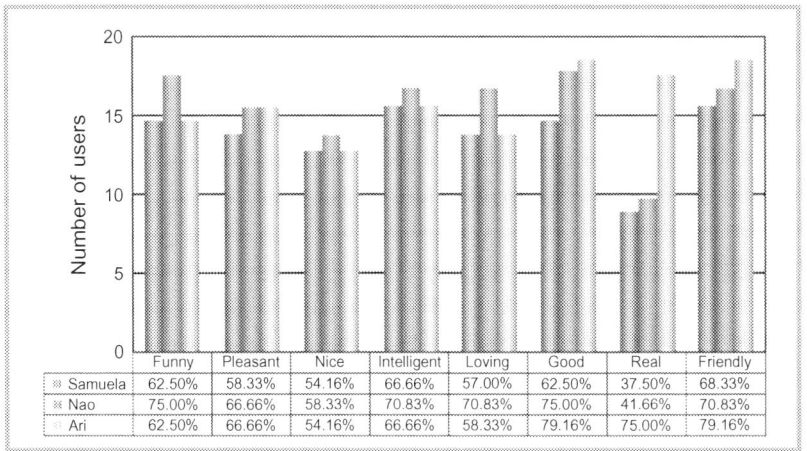

Figure 7.13 Descriptors qualifying each Companion's personality. *Courtesy of Dr. Néna Roa Seïler, unpublished doctoral dissertation.*

accurate translation into English is probably pleasant, a concept that also baffled the children and defied definition even though it was proposed by one of them.

Loving applied to persons' actions of emotional characteristics—"A loving person is somebody who hugs you and somebody who speaks nicely to you"; "he is affectionate because he/she buys things for me," "she does what I say," and to another, "a person who feels compassion for children and helps them." The interviews identified Ari as the most affectionate because of her behavior—"She smiled," "She talked nice." Nao, not having a mouth couldn't smile which generated some confusion; nevertheless children found him fascinating probably because none of them had met a robot before, whereas they had already encountered screen-based characters in video games.

PHYSICAL DESCRIPTORS

Figure 7.14 presents some of the Adjectives, descriptors used to describe the physical appearance of Companions. Some children had no clear idea of the concept of beautiful proposed by one of them and some found it difficult to define it, but with the help of the SD, Nao the virtual tutor was picked out by 17 children as the most beautiful and as having an attractive physical appearance, 16 opted for Ari, and 15 for Samuela, that is, all three were judged to be almost equally attractive. Serious was a concept associated with someone who is lazy or does not like to work—"a serious person is one who misbehaves,

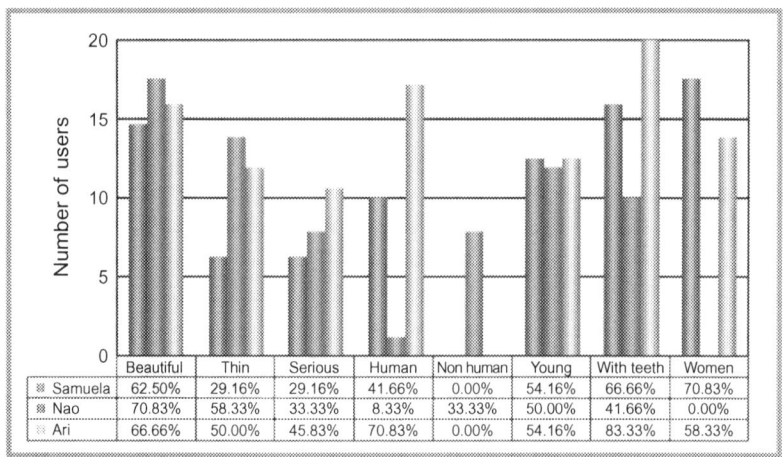

Figure 7.14 Descriptors qualifying physical appearance. *Courtesy of Dr. Néna Roa Seïler, unpublished doctoral dissertation.*

someone who does not smile, speaks or plays." It was also associated with facial expressions. Ari was found to be the most serious, followed by Nao, then Samuela, the latter judged so because she didn't smile very often. Ari and Samuela looked human based on physical appearance, on anthropomorphic features, whereas Nao was seen as nonhuman because he was a robot, seen as not being alive. Children viewed young mostly in terms of what grade they were in school, then how people behaved, how they dressed. To them – "you're young if you are in Secondary school" as against "children" if you are still in Primary school. Nao was seen as the youngest, they considered him as a child like themselves very likely because of his size. "With teeth" or "having teeth" was considered a sign of youth, as in our culture wrinkles are that of old age. Twenty children rated Ari as "having teeth." In contrast, Nao, whom children regarded as being similar to them, was rated lower. Children considered real those things that they could see and touch and unreal a product of the imagination—"witches are unreal." These distinctions failed in practice and Ari and Samuela were determined as real because they looked real, while Nao looked unreal because he was a robot.

UTILITY OR FUNCTIONALITY DESCRIPTORS

Figure 7.15 presents the Companion's utility descriptors. Here the results were very clear. The children defined the utility of the Companion as a provider of help, one that offers knowledge, therefore highly beneficial and

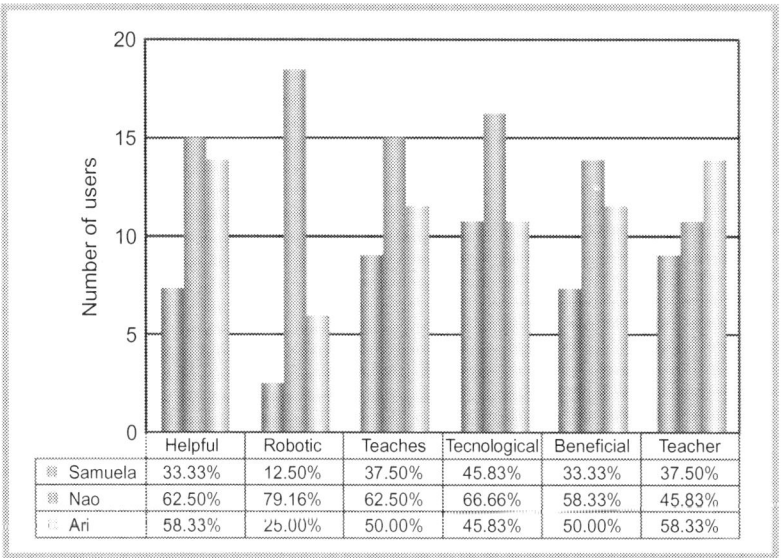

Figure 7.15 Descriptors describing Companions' utility. *Courtesy of Dr. Néna Roa Seïler, unpublished doctoral dissertation.*

valuable. Although all Companions were rated as helpful, here we see evidence of the "novelty effect" at play, designating Nao as more helpful and beneficial than the others.

INTERACTION WITH A COMPANION

Table 7.9 presents the answers to the question—"How would you like your interaction with the Companion to be?"—and the type of interaction children preferred overall was in the affective domain: be more loving, smile more, and look happier and that it would speak nicely like a friend, as the phrases highlighted in Table 7.9.

Table 7.9 Interaction based on behavior expected from Companions

Samuela	Nao	Ari
Smile more	Move more	Be more loving
Talk more	Let himself be touched	Be more like a teacher or a friend
Talk like a teacher or a friend	Be more helpful	Smile more
	Be like a friend	Be happier
	Talk more like a child	

Courtesy of Dr. Néna Roa Seïler, unpublished doctoral dissertation.

PREFERRED ACTIVITIES WITH COMPANIONS

Table 7.10 shows the activities that children chose to perform with each Companion, ludic activities related to learning being the top choice—Samuela teaching them how to play with the computer and Ari to recite poems because Samuela and Nao had flat voices without intonation. Samuela was asked to talk about herself and her life, indicative of a need for companionship. There was also expectation that Ari and Samuela speak like a teacher, representing the protective adult in children's mind.

Nao had the robotic "novelty effect", yet was perceived as a child and expected to speak like a child, perform activities such as dance, take a walk or play, and also to sing with them, all signs of their longing for companionship. Children's perception sways toward the human side of the agent (teach me, play with me, tell me stories, recite poetry, etc.) and the machine side is more blurred (e.g., they believe that Companions possess knowledge the way teachers or adults do, rather than be able to access, gain knowledge from the Internet).

PREFERRED COMPANION FOR SUPPORTING DIFFERENT SUBJECTS

Mathematics, geography, natural sciences, civics, ethics, and Spanish language were the subjects in which the children felt they most needed support from the Companion. Of the 24 pupils, 9 selected Ari as support because she was more talkative and could therefore explain better, this again underlines the fact that children recognize codes of human communication without knowing how. Seven chose Nao for her "novelty" and on considering him to be the most intelligent and five said that any Companion would be helpful. Three children chose Samuela because "she was the cleverest

Table 7.10 Activities children would like to perform with Companion

Samuela	Nao	Ari
Play with them on the computer.	Play with them outdoors: run, play football, play basket ball.	Recite poems.
Tell stories or horror stories.	Teach them to dance.	Show them videos.
Talk about herself and her life.	Teach them to sing.	Tell stories.
Listen to them.	Allow himself to be touched.	

Courtesy of Dr. Néna Roa Seïler, unpublished doctoral dissertation.

of all the tutors." All the children chose at least one Companion for each subject, proof that they could or would rely on them to help them with their learning.

SIGNIFICANCE OF COMPANIONS

In order to ascertain the perception of and the feeling toward the Companions in general, the children were asked the question—"What is a Companion for you?" Most, 80% of the children, regarded the Companion as a friend, 10% as a teacher, and the other 10% as a teacher's assistant (Figure 7.16). Each Companion was referred to by its name—Samuela, Nao, or Ari—rather than the word "Companion".

TRUST IN COMPANIONS

To the question—"Do you trust Companions?" —*all* answered "Yes". Trust makes learning any subject with a Companion possible, speaking with them about their daily lives, sharing their secrets and their problems at home, not feeling ashamed to ask them questions. It seems that Companions are perceived to have trustworthiness, a genuine attribute that facilitates the developing of friendship and companionship, thus an important factor in designing systems for children.

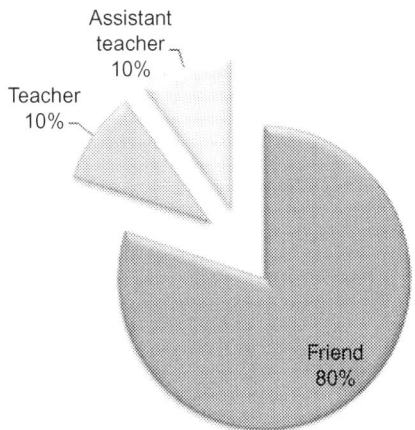

Figure 7.16 What Companions signify to children.

EMOTIONS TOWARD COMPANIONS

The prEmo tool (Premotool, 2013) was used for measuring emotions elicited also by the three Companions which were in general positive, joy, admiration, fascination, and desire featuring strongest, as shown in Figure 7.17.

Only two children reported negative emotions, one finding interaction with Ari boring perhaps because of not understanding the rules of the game properly, that the Companion is there to help find, not to give the answers. Another child, not having met a robot before mentioned feeling fear when meeting Nao. Oddly, Nao scored well on eliciting admiration (25%), fascination (20.83%), and desire (20.83%), the icon proposed by prEmo (Premotool, 2013) showing in Figure 7.18 the kind of desire aroused in the children by Nao.

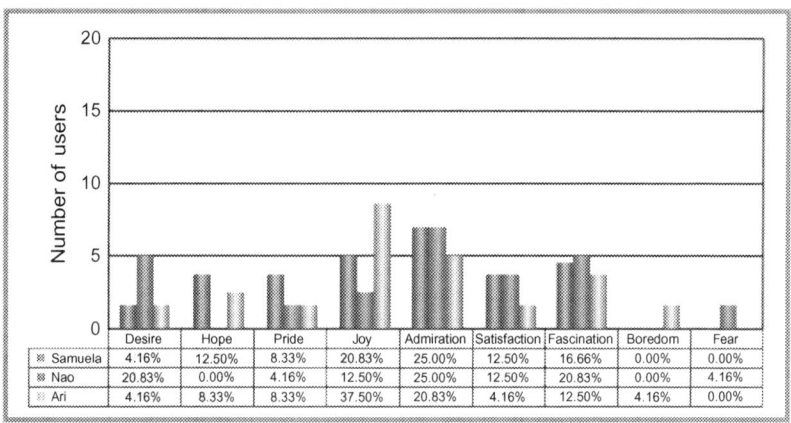

Figure 7.17 Emotion felt by children toward the three Companions. *Courtesy of Dr. Néna Roa Seïler, unpublished doctoral dissertation.*

Figure 7.18 The prEmo icon chosen by children to express their feeling toward Nao. *Courtesy of Susagroup (Premotool, 2013).*

This icon of a character with extended arms matched their feelings of really wanting physical contact with, to touch Nao, confirming a strong desire for physical activities with it—going for a walk, playing football, or being taught how to dance—as already shown in Table 7.10. This finding indicates that although children feel comfortable with screen based Companions and express strong emotions toward them, embodiment impacts powerfully on the kind of activities they would like to do with any particular Companion.

Joy, a basic emotion of great pleasure and happiness that features in the Ekman (1999) and the Plutchik (2001) Models of Emotions, adds to developing trust in and enjoyment of learning together and admiration is a particular form of positive affect that motivates self-improvement (Algoe & Haidt, 2009). Impressed by the amazing things an ordinary Companion could do—giving advice during the game, comforting the children when they faltered, and organizing the players' turns—children quickly came to trust it to help their progress and looked to have a long term relationship with it.

The above confirm that children perceive these entities to be human or human-like and that their presence and interaction with them arouse positive emotions in them, based largely on their judging body gestures in the case of Nao or the facial expressions of Ari and Samuela to be genuine.

DISCUSSION OF RESULTS

Children interacting in real time with three different Companions, each in its own embodiment, applying three different categories of Interaction Strategies and the WoZ method generated a natural context in which their constructing social relationships with Companions was easily observable and more importantly for our purposes, by playing a "Serious Game," the measurement of their feelings toward these virtual Companions was greatly facilitated.

The most notable finding, its undeniable factual validity also constituting a discovery is that children perceived Companions in positive terms and that these entities evoked positive emotions in them, facilitated companionship and friendship, and that they all share this capacity. The strong relationship between affect and cognition are highly relevant to the learning experience as positive emotions are conducive to better processing complex data, receiving new information and accepting challenges (Estrada, Isen, & Young, 1997; Raver, 2003). Aspects of positive emotions are a reflection of happiness and wellbeing that also project into the future (Fredrickson & Joiner, 2002), engender the complex understanding of the process of people forming new relationships, motivate self-improvement and nurture successful intergroup relationships (Reeves & Nass, 1996, p. 153), attributes which, as it is the finding of this paper, make them ideal candidates for use in settings where care, encouragement, and empathy are priorities.

All three Companions attracted the children's attention in different ways and none was rejected; in fact all were rated well as being fun, pleasant, nice, intelligent, loving, good, and friendly with Nao scoring marginally higher in all these aspects. It was clear that children wanted affective interaction with more smiling, talking, chatting like friends, meeting in the breaks or after classes, and that their overriding need was for companionship and that these entities were not to be as Teachers (authority figures).

Children evaluated Companions relative to themselves—"we are children because we attend Primary School, those attending Secondary School are adults"—regarding Nao as a child on that basis, and it would make an interesting study to see how applying this perspective to measure friendship and companionship would play out in a scenario of a "Child- Screen based Companion" interacting with children or indeed with a male Companion.

As to activities preferred by children, their choices for a screen-based Companion differed from that for a robot. Personalization promotes engagement, as when children were noticeably affected by Samuela calling them by their names or mentioning some detail from a previous encounter to the nine children who had met her in a previous test.

The results of the Questionnaires showed that children found defining certain Adjectives challenging and often provided Nouns instead, confirming the superiority of nonverbal tools for measuring children's feelings. It also showed up the disparity in the respective semantic knowledge of urban versus rural schoolchildren, that in rural schools children of mixed abilities all learn together. It also became evident that their definitions were based primarily on visual perception, appearance, observed behavior, and actions.

The powerful role of the "willing suspension of disbelief" (Englis, 1992; Bates, 1994)—a paradox, that users are aware of the Companion not being human, nevertheless they are willing to go along with the illusion that they are, even invent the kind of relationship they want along the human-object relationship spectrum, which supports the view that computers are social actors (Reeves & Nass, 1996)—is another important outcome. Children would have encountered similar synthetic agents in videogames or on television and knew these to be nonsentient, nevertheless they were happy to *play along* and freely share their personal details and opinions with them.

SUMMARY AND FUTURE WORK

The purpose of this research was to explore the design and implementation challenges in interaction between Companions endowed with emotional abilities and children, in a learning environment, and in real time. Reviewing existing know-how and technology indicated user expectation for social, affective relationship with utility of real benefit and the experiments described here examined specifically how children perceived and experienced artificial entities. The protocol WoZ was developed to create and test Conversational, Empathetic, and Domain Specific Interaction Strategies to deliver engaging, affective, long term relationship so important to children's healthy emotional and social development. The AC was designed as the resource of interactive capabilities that made the WoZ, emulating the human communication model, able to implement the learning "Serious Game." At the heart of this work is children learning by playing, Companions playing with them, rather than acting in a pedagogical capacity. Results highlighted the desirability of companionship and trust in Companions and their acceptance as attractive, likeable, and that their impact is positive and beneficial in many ways.

In a future work the WoZ technique with Interaction Strategies presented here will be used to investigate improvement in learning and playing games and in other children's activities that could be shared with Companions.

REFERENCES

Adam, C., Cavedon, L., & Padgham, L. (2010). Hello Emily, how are you today?: Personalised dialogue in a toy to engage children. In *Presented at the proceedings of the 2010 workshop on companionable dialogue systems, association for computational linguistics* (pp. 19–24). USA: ACL.

Algoe, S. B., & Haidt, J. (2009). Witnessing excellence in action: The "other- praising" emotions of elevation, gratitude, and admiration. *The Journal of Positive Psychology*, 4(2), 105–127.

Alvarez, J. (2007). Du jeu vidéo au serious game: approches culturelle, pragmatique et formelle. Doctoral dissertation, Université de Toulouse II (Paul Sabatier).

Arroyo, I., Woolf, B. P., Royer, J. M., & Tai, M. (2009). Affective gendered learning companions. In *Proceedings of the 2009 conference on artificial intelligence in education: Building learning systems that care: From knowledge representation to affective modelling* (pp. 41–48). Amsterdam, Netherlands: IOS Press.

Barab, S., Dodge, T., Tuzun, H., Job-Sluder, K., Jackson, C., Arici, A., et al. (2007). The quest atlantis project: A socially-responsive play space for learning. In B. E. Shelton & D. Wiley (Eds.), *The educational design and use of simulation computer games* (pp. 159–186). Rotterdam, Netherlands: Sense Publisher.

Barab, S., Thomas, M., Dodge, T., Carteaux, R., & Tuzun, H. (2005). Making learning fun: Quest Atlantis, a game without guns. *Educational Technology Research and Development*, 53 (1), 86–107.

Bates, J. (1994). The role of emotion in believable agents. *Communications of the ACM, 37*(7), 122–125.

Benyon, D., Gamback, B., Hansen, P., Mival, O., & Webb, N. (2013). How was your Day? evaluating a conversational companion. *IEEE Transactions on Affective Computing, 4*(3), 299–311.

Benyon, D., & Mival, O. (2013). Scenarios for companions. In *Your virtual butler* (pp. 79–96). Berlin, Heidelberg: Springer-Verlag.

Bickmore, T. W., & Picard, R. W. (2005). Establishing and maintaining long-term human-computer relationships. *ACM Transactions on Computer-Human Interaction (TOCHI), 12* (2), 293–327.

Bradley, J., Mival, O., & Benyon, D. (2009). Wizard of oz experiments for companions. In *Presented at the proceedings of the 23rd British HCI group annual conference on people and computers: celebrating people and technology, British computer society* (pp. 313–317).

Busso, C., & Narayanan, S. (2008). Recording audio-visual emotional databases from actors: A closer look. In *Second international workshop on emotion: corpora for research on emotion and affect, international conference on language resources and evaluation (LREC 2008)* (pp. 17–22).

Cassel, J., Sullivan, J., Prevost, S., & Churchill, E. (Eds.), (2000). *Embodied conversational agents*. Boston: MIT Press.

Chiu, M. M. (2000). Group problem-solving processes: Social interactions and individual actions. *Journal for the Theory of Social Behaviour, 30*(1), 26–49.

Craig, P., Roa-Seïler, N., Lara Rosano, F., & Martínez Díaz, M. (2013). The role of embodied conversational agents in collaborative face to face computer supported learning games. In *Proceedings of the 26th international conference on system research, informatics and cybernetics, Baden, Germany*.

De Sevin, E., Niewiadomski, R., Bevacqua, E., Pez, A.-M., Mancini, M., & Pelachaud, C. (2010). Greta, une plateforme d'agent conversationnel expressif et interactif. *Technique Et Science Informatiques, 29*(7), 751.

Desmet, P. (2005). Measuring emotion: Development and application of an instrument to measure emotional responses to products. In M. A. Blythe, A. F. Monk, K. Overbeeke, & P. C. Wright (Eds.), *Funology: From usability to enjoyment* (pp. 111–123). Dordrecht: Kluwer Academic Publishers.

Ekman, P. (1999). Facial expressions. In T. Dalgleish (Ed.), *Power handbook of cognition and emotion:16*. Sussex, UK: John Wiley & Sons, Ltd. pp. 301–320.

Elias, M. J. (1997). *Promoting social and emotional learning: Guidelines for educators*. Alexandria, VA: ASCD.

Englis, B. G. (1992). The willing suspension of disbelief and its importance in understanding advertising effects. *Marketing Theory and Applications, 3*, 203.

Estrada, C. A., Isen, A. M., & Young, M. J. (1997). Positive affect facilitates integration of information and decreases anchoring in reasoning among physicians. *Organizational Behavior and Human Decision Processes, 72*(1), 117–135.

Fredrickson, B. L., & Joiner, T. (2002). Positive emotions trigger upward spirals toward emotional well-being. *Psychological Science, 13*(2), 172–175.

Heylen, D., op den Akker, R., ter Maat, M., Petta, P., Rank, S., Reidsma, D., et al. (2011). On the nature of engineering social artificial companions. *Applied Artificial Intelligence, 25* (6), 549–574.

Hsu, S. H., Chuang, M. C., & Chang, C. C. (2000). A semantic differential study of designers' and users' product form perception. *International Journal of Industrial Ergonomics, 25*(4), 375–391.

Keltner, D., Ekman, P., Gonzaga, G. C., & Beer, J. (2003). Facial expression of emotion. In R. J. Davidson, K. R. Scherer, & H. H. Goldsmith (Eds.), *Series in affective science. Handbook of affective sciences* (pp. 415–432). New York, NY: Oxford University Press.

Lakin, J. L., & Chartrand, T. L. (2003). Using nonconscious behavioral mimicry to create affiliation and rapport. *Psychological Science*, *14*(4), 334–339.

Lee, J. K., Stiehl, W. D., Toscano, R. L., & Breazeal, C. (2009). Semi-autonomous robot avatar as a medium for family communication and education. *Advanced Robotics*, *23*(14), 1925–1949.

Leite, I., Mascarenhas, S., Pereira, A., Martinho, C., Prada, R., & Paiva, A. (2010). Why can't we be friends?" An empathic game companion for long-term interaction. In *Proceedings of the 10th international conference on intelligent virtual agents* (pp. 315–321). Berlin, Heidelberg: Springer.

Lim, M. Y. (2012). Memory models for intelligent social companions. *Human-computer interaction: The agency perspective* (pp. 241–262). Berlin, Heidelberg: Springer-Verlag.

Lopez Mencia, B. (2011). Agentes animados personificados en sistemas interactivos: diseño y evaluación. Doctoral thesis, Universidad Politecnica de Madrid.

Maatman, R. M., Gratch, J., & Marsella, S. (2005). Natural behavior of a listening agent. In T. Panayiotopoulos, J. Gratch, R. S. Aylett, D. Ballin, P. Olivier, & T. Rist (Eds.), *Intelligent virtual agents* (pp. 25–36). Berlin, Heidelberg: Springer.

Majors, K. A. (2013). Children's perceptions of their imaginary companions and the purposes they serve: An exploratory study in the united kingdom. *Childhood*, *20*(4), 550–565.

Markopoulos, P., Read, J. C., MacFarlane, S., & Hoysniemi, J. (2008). *Evaluating children's interactive products: Principles and practices for interaction designers*. San Francisco, CA: Morgan Kaufmann.

Martínez García, D., Craig, P., Roa-Seïler, N., & Benítez Saucedo, A. (2012). Validación de una estrategia de interacción de un agente corpóreo conversacional a través de la técnica del mago de Oz. In *Proceedings of MexIHC, Mexico City, Mexico*.

Maslow, A. H. (1943). A theory of human motivation. *Psychological Review*, *50*(4), 370.

McConnell, A. R., Brown, C. M., Shoda, T. M., Stayton, L. E., & Martin, C. E. (2011). Friends with benefits: On the positive consequences of pet ownership. *Journal of Personality and Social Psychology*, *101*(6), 1239.

Mouaheb, H., Fahli, A., Moussetad, M., & Eljamali, S. (2012). The serious game: What educational benefits? *Procedia-Social and Behavioral Sciences*, *46*, 5502–5508.

Muñoz, K., Mc Kevitt, P., Lunney, T., Noguez, J., & Neri, L. (2010). PlayPhysics: An emotional games learning environment for teaching physics. In *Knowledge science, engineering and management* (pp. 400–411). Berlin, Heidelberg: Springer.

Plutchik, R. (2001). Integration, differentiation, and derivatives of emotion. *Evolution and Cognition*, *7*(2), 114–125.

Premotool. (2013). *Courtesy of Susagroup, In prEmo, measure product emotions*. Retrieved June 26, 2013, from http://www.premotool.com/

Quest Atlantis (2014), extracted from http://atlantisremixed.org/on March 2014.

Raskin, N. J., Rogers, C., & Witty, M. C. (2007). Client-centered therapy. In R. J. Corsini & D. Wedding (Eds.), *Current psychotherapies* (pp. 141–186). CA: Thomson Higher Education Belmont.

Raver, C. (2003). Young children's emotional development and school readiness. *Social Policy Report*, *16*(3), 3–19.

Reeves, B., & Nass, C. (1996). *How people treat computers, television, and new media like real people and places*. New York: CSLI Publications and Cambridge University Press.

Roa Seïler, N. (2015). Towards an Emotionally Intelligent Interaction Strategy for Multimodal ECAs Acting as Companions. Doctoral dissertation, Edinburgh Napier University.

Roa-Seïler, N., Benyon, D., & Leplâtre, G. (2009). An affective chanel for companions. In *Electronic proceedings empathic agents workshop at AAMAS, Proc. of 8th int. conf. on autonomous agents and multiagent systems, Budapest, Hungary*.

Roa-Seïler, N., Craig, P., Aguilar, J. A., Benítez Saucedo, A., Martínez Díaz, M., & Lara Rosano, F. (2014). Defining a child's conceptualization of a virtual learning companion. In *proceedings of international technology, education and development conference* INTED, Valencia, Spain. ISBN: 978-84-616-8412-0.
Rogers, C. (1995). *A way of being*. Boston, MA: Houghton Mifflin Harcourt.
Schröder, M. (2006). Perception of non-verbal emotional listener feedback. In *Proceedings of SPEECH PROSODY, Dresden.* .
Shaw, C., Brady, L. M., & Davey, C. (2011). *Guidelines for research with children and young people*. London: National Children's Bureau Research Centre.
Smith, C., Crook, N., Dobnik, S., Charlton, D., Boye, J., Pulman, S., et al. (2011). Interaction strategies for an affective conversational agent. *Presence: Teleoperators and Virtual Environments, 20*(5), 395–411.
Somsai, S., & Intaraprasert, C. (2011). Strategies for coping with face-to-face oral communication problems employed by Thai university students majoring in English. *GEMA Online™ Journal of Language Studies, 11*(3), 83–96.
Stiehl, W. D., Lee, J. K., Breazeal, C., Nalin, M., Morandi, A., & Sanna, A. (2009). The huggable: A platform for research in robotic companions for pediatric care. In *Presented at the proceedings of the 8th international conference on interaction design and children* (pp. 317–320): ACM.
The British Association for the person centred approach extracted October 2013 from http://www.bapca.org.uk/about/what-is-it.html.
Ten Dam, G., & Volman, M. (2007). Educating for adulthood or for citizenship: Social competence as an educational goal. *European Journal of Education, 42*(2), 281–298.
Thiebaux, M., Lance, B., & Marsella, S. (2009). Real-time expressive gaze animation for virtual humans. In *Presented at the proceedings of the 8th international conference on autonomous agents and multiagent systems-volume 1, International Foundation for Autonomous Agents and Multiagent Systems* (pp. 321–328).
Vosniadou, S. (2002). Mental models in conceptual development. In L. Magnani & N. Nersessian (Eds.), *Model-based reasoning* (pp. 353–368). New York: Springer.
Wilks, Y. (2005). Artificial companions. In *Machine learning for multimodal interaction* (pp. 36–45). Berlin, Heidelberg: Springer-Verlag.
Wilks, Y. (2006). *Artificial companions as a new kind of interface to the future internet*. Oxford: Oxford Internet Institute.
Wilks, Y. (2010). Is a companion a distinctive kind of relationship with a machine? In *Presented at the proceedings of the 2010 workshop on companionable dialogue systems, association for computational linguistics* (pp. 13–18).

SECTION II

Critical Theoretical Engagements with Emotions, Technology, and Design

CHAPTER 8

The Emulation of Emotions in Artificial Intelligence: Another Step into Anthropomorphism

Mariana Goya-Martinez
University of Illinois at Urbana-Champaign, Champaign, IL, USA

INTRODUCTION: REDEFINING THE HUMAN

Scientific discourses within artificial intelligence, cognitive science, neurology, and genetics promote a new definition of human as a replicable and predetermined entity. Attempts to simulate and duplicate human characteristics—from physical appearance and behavior to mental processes and emotions on robots as well as on software agents—exemplify how the belief in the replicability of human nature exists within certain branches of artificial intelligence. Within cognitive science, connectionism conceptualizes the human mind as an information processing machine whose laws and rules are written on the brain. In neurology, advances in brain imaging techniques promise to answer all the questions regarding personality, emotions, and the identity of an individual by looking at a person's brain. Some areas of genetics try to partially answer these questions by studying the DNA. Roboticists, neuroscientists, and computer scientists have been working together in several projects—like the Human Brain Project—with the objective of simulating the brain, developing interactive supercomputing, and developing brain-inspired computing and robotics (Human Brain Project, 2015). These kinds of projects reinforce the notion that the mysteries of the brain can and will be fully understood someday and, once this happens and the appropriate technology has been developed, humans could be easily replicated. It is no coincidence that this trend has a special relationship with technology; some researchers in artificial intelligence and cognitive science use their own computer programs as evidence to support their beliefs, whereas in neurology and genetics, sophisticated technology gives them access to the brain and DNA. The specific discourses that sponsor the notion of humans as replicable entities within cognitive science, neurology, and genetics are beyond the scope of the present research. The emulation of emotions in machines, which is the focus of this paper, belongs to this broad philosophical trend promoting a higher level of anthropomorphism within the field artificial intelligence.

Since the beginning of the field of artificial intelligence, inventors created and designed computers, not only to make numerical operations and information handling easier, but also by anthropomorphic motivations. In 1911, Torres y Quevedo created the first automaton capable of replacing a human chess player (Randell, 1982). Since then, computer inventors have been trying to emulate and implement certain human characteristics in machines, from problem-solving skills, decision-making processes, and the use of language to turn-taking gestures and, more recently, emotions. While contemporary computer scientists are trying to show that their creations are human-like, they do not explain the reasons behind the emulation of human behavior and human internal processes in their intelligent agents.

In the technological race toward the creation of an intelligent and emotional computer agent, several questions emerge. For instance, is there something unique about humans? What human characteristics could or should never be emulated by machines? How does the emulation of human characteristics, especially emotions, redefine the human? What are the ideas and motivations behind the emulation of emotions? The ideas of today's inventors will shape the technologies that humans will use tomorrow. Since technologies are not neutral, the ideas they embody can influence the life of their users and their societies. In this case, the use of words such as "emotions," "beliefs," "autonomy," "consciousness," "intelligence," and "thinking" in order to describe the capabilities and features of machines, not only distorts their meaning, but also anthropomorphizes machines, ultimately redefining what being human means. Therefore, artificial intelligence redefines the concept of "human"; an entity that can be replicated and compared to a highly advanced intelligent machine.

To analyze the motivations behind the emulation of emotions on machines, I use the ideas of artificial intelligence researchers that were either published or expressed in public or personal interviews (Goya-Martínez, 2008). The objective of this research is to study the redefinition of emotions within artificial intelligence, the motivations behind their emulation in machines, and how this contributes to a higher level of anthropomorphism within the field. First, I present how the definition of emotion has evolved within the field of artificial intelligence. I also examine the implications of these definitions. Then, I introduce the differences between strong and weak artificial intelligence and how these two branches try to emulate emotions in their agents' designs. After this, I explore the motivations of computer scientists behind the emulation of emotions in artificial intelligence. Finally, I relate this phenomenon with anthropomorphism and I suggest some possible advantages and disadvantages of this emulation.

THE DEFINITION OF EMOTIONS IN ARTIFICIAL INTELLIGENCE

The separation of reason from emotions and feelings has been present in the field of artificial intelligence since its origins (Newell, 1982). While emotions were once considered as human flaws that make us subjective and irrational, they have been redefined as part of human intelligence and, thus, in the case of higher emotions, are now being used to differentiate humans from animals (Kurzweil, 2012; Newell, 1982). Following the path of information theory, higher emotions, as other cognitive percepts, are considered information patterns that can circulate from one material substrate to another and remain unchanged. Therefore, emotions, as other human characteristics such as language and beliefs, are redefined as replicable in an intelligent agent's design.

Influenced by popular Christian folk psychology, machines were once considered within artificial intelligence more rational than humans because they could perform "cold logic," with no feelings or emotions conflicting with their reason (Newell, 1982). Emotions were not considered useful, i.e., rational decision-making, mental functions, like learning or thinking. Many considered the idea of artificially replicating emotions—if it were possible—a dangerous one, believing that such action could result in an irrational and dangerous monster or machine, just as terrifying as Dr. Frankenstein's creation (Shelley, 1818). Newell's account of the history of the field shows that this dichotomy was still present in 1970s, when emotions and feelings were not considered necessary for the development of intelligent machines and they were regarded as "program-resistant functions" that could not be mechanized (Newell, 1982, p. 28).

The definition of emotions has evolved within artificial intelligence, inspired by the fields of neurology and cognitive science, which define emotions as part of human intelligence. This interest in neurology and cognitive science is explicitly present in the books of many famous researchers of artificial intelligence and is used as evidence of the replicability of human intelligence, in which emotions are now considered a necessary ingredient (Kurzweil, 1999, 2012; Minsky, 1988, 2006). Minsky, one of the founders of MIT's artificial intelligence program, explains that it is useless to separate intellectual from emotional mental processes since both are "ways of thinking" (Minsky, 2006). The human brain focuses on one way of thinking and, if this is not achieving optimal results, it switches to another. These different ways of thinking are conceptualized as instincts, similar to hunger and thirst, which are the result of evolutionary processes that fulfill certain objectives for the survival of the individual and of the species. For instance, falling in love turns off resources, like critical analysis, in order to make the individual unaware of the defects of the loved person. Similarly, anger turns on certain resources that allow the individual to react unusually strong and fast to certain events, while it turns off other resources that make the individual act more prudently and critically. For Minsky (2006), emotions, as any other way of thinking, are part of our intelligence and the result of neural network processes within our brains.

Kurzweil (2012), current director of Google's artificial intelligence engineering department, distinguishes two kinds of emotions: lower emotions, produced by our "old brain" inherited from reptiles, and, higher emotions, produced by the mammal neocortex which finds it most evolved stage in the human brain. Therefore, although lower emotions still try to set the agenda, the neocortex in the human brain controls the outputs of the amygdala—a structure inherited from primitive brains—with the prefrontal cortex, one of the decision-making regions of the "new brain." For Kurzweil, in order to emulate human behavior, modeling emotions coming from both the "old" and the "new" brain would be completely necessary. However, to emulate the human cognitive experience, it would be necessary to model only the higher-level emotional outputs produced by the neocortex. Higher-level emotions in the new brain are largely the result of neural connections. For instance, spindle neurons, which are recently evolved brain structures, deeply interconnect different regions of the neocortex and are particularly active when individuals experience higher-level emotions such as when they hear their baby cry or when they listen to music they enjoy (Kurzweil, 2012). These neurons, which are the largest in the human brain, show that emotions are interconnected with perceptual and cognitive regions and participate in decision-making processes and moral judgments (Kurzweil, 2012). Following Minsky, Kurzweil (2012) considers that higher emotions are represented in neural connections as any other cognitive precept and, therefore, replicating them should be possible once the human brain is scanned in all its intricacy.

Even though there still is a certain dichotomy between reason and emotion, I found similar discourses when I engaged in qualitative research by interviewing several artificial intelligence researchers from two American Midwestern

Universities (for more details about the study see Goya-Martínez, 2008). In that study, I used an IRB protocol that ensured participants total anonymity so that they could be totally honest in their answers to my questions (IRB 2006-0150, University of Illinois at Urbana-Champaign). The researchers interviewed described computers as intelligent, rational, and autonomous agents, capable of learning. Intelligence was defined as a computer's capacity to react to the environment. Regarding rationality, it could be inferred from their statements that their notions derive from economic theories and rational choice models that argue that rational behavior is motivated by a conscious calculation of advantages, based on an explicit and consistent value system (Schelling, 1960). In this sense, these theories presuppose that humans make decisions based on rationality and common knowledge. For these researchers, however, humans were not defined as completely rational, although capable of making rational choices and because of this these rational models apply to both: humans and machines. Similar to the mainstream discourses in the field, emotions were defined as instincts that aided human intelligence since they were used as "shortcuts" for thinking and decision making. Some of the researchers also used the discourse of evolution to explain why emotions in humans were necessary for their survival. Many believed that the implementation of emotions in machines might allow the construction of a coherent system that is able to have a personality and to efficiently handle several diverse tasks at the same time. However, some of their responses showed that, even when mainstream discourses disregard the idea of emotions as flaws, the dichotomy reason-emotion has not left the field completely. The same researchers interviewed who conceptualized emotions as instincts thought that emotions made humans less perfect than machines since they make them subjective. Therefore, machines were interpreted as more rational than humans because the latter have prejudices and emotions. One researcher exemplified this by saying that machines are better designed for decision-making in dangerous situations—like throwing a bomb in a war—precisely because they do not have feelings, emotions, and/or personal interests. All the researchers interviewed agreed that a machine could not be irrational and that if it shows irrational behavior it would be because it is rationally following a program designed to make its behavior appear as irrational.

Since many researchers inside the field consider emotions to be represented in neural networks somewhat underestimating the fact that these are also based on chemical reactions, emotions were also redefined as information patterns. If emotions are information patterns, then they can be artificially generated. This idea follows the trend that considers the human mind as information patterns written in the brain. This trend started with Turing's imitation test and Shannon's and Wiener's information theory (Hayles, 2008). On the one hand, Turing (1999) proposed the "imitation game" to test the human-likeness of digital computers. Previously, the imitation game tested an interrogator's ability to distinguish who was a man and who was a woman between two players. The interrogator was in a separate room and could only have typewritten communication with the other two players. The man would simulate a woman and the woman would perform as herself. After asking several questions, the interrogator would guess the gender of both players. To apply this game to digital computers, Turing proposed to exchange the male player for a machine. Therefore, the interrogator would not guess players' gender but about their humanness. He wanted to draw a line "between the physical and intellectual capacities of a man" (Turing, 1999, p. 38). Hence, Turing's test redefines humanness, and other human characteristics such as intelligence and emotions, as information present in the discourse and not in the psychological attributes of the player.

On the other hand, Shannon separates information from the substrate that carries it (Hayles, 2008). In Wiener's words: "information is information, not matter or energy," thus it remains unchanged when it moves from one medium to another (1965, p. 132). Following this idea, Kurzweil (1999) and Moravec (1988) propose to replicate a human by downloading or scanning information from the brain into a computer. Therefore, emotions, personality, and the rationality of an individual are regarded as information or data that can be transferable to another substrate and remain unchanged. Minsky states that: "The most important thing about each person is the data and the programs in the data that are in the brain" (quoted in Hayles, 2008, p. 245). The reduction of emotions as information patterns contradicts the thesis of some neurologists like Damasio (2005), in which the body plays an important role for the mind. It also opposes McLuhan's maxim "The medium is the message" (McLuhan & Fiore, 1967), where content changes if it is transferred to a different medium since each medium has different sensory and intellectual biases.

When emotions were considered as flaws, artificial intelligence researchers were not interested in replicating them. Redefining emotion as a cognitive percept and as necessary to reproduce human cognitive abilities makes its emulation an objective within the field. This emulation closes the gap between the definition of human and intelligent machines a little further, since emotion is no longer separated from reason and is now considered to be replicable in artificial neural networks. While emotion and reason remain as different concepts, they are also considered to work together in human intelligence and decision making processes. In spite of their differences, both emotion and reason have been reconceptualized as information patterns that can be enacted in the brain as well as in a computer. Not only are machines considered to be able to have emotions, but also humans are considered to be machines that process information. "The Emotion Machine" and "The Age of Spiritual Machines" written by Minsky (2006) and Kurzweil (1999), respectively, illustrate these comparisons in their titles since, and by following their arguments, the metaphor could apply to humans as well as to computers.

THE ROLE OF EMOTIONS IN INTELLIGENT AGENTS' DESIGN
Statistical Versus Simulation Mode, Strong Versus Weak Artificial Intelligence

Artificial intelligence researchers use two strategies, or their combination, to try to emulate human characteristics in their designs: performance or statistical and cognitive or simulation approaches. These two techniques correspond to two different positions within artificial intelligence: strong and weak. One of their main differences resides in their definition of replication. In both positions, however, the emulation of emotions plays an important objective within their designs.

There are two ways to try to emulate humans' capacities: performance and simulation mode (Weizenbaum, 1976). The performance or statistical mode tries to emulate human characteristics by applying the most efficient principle or rule. The other trend, cognitive or simulation mode, tries to achieve humans' performance through imitating human mental processes; it wants to achieve human-like behavior only through the use of rules and theories that apply to humans. This can be easily illustrated through the techniques employed in the emulation of natural language. The objective of this branch of artificial intelligence, natural language processing, is to develop discourse in computer agents. The statistical approach consists in the utilization of empirical data, such as recorded conversations, to attach certain meaning to words according to probabilistic rules and to produce statistically coherent responses. By contrast, the cognitive or simulation approach operates within linguistic and psychological frameworks, such as semantic analysis and knowledge representation. This technique tries to emulate language processing based on grammatical and semantic theories. The performance mode's goal is to build machines that appear intelligent, which is more modest than the simulation mode's goal. The cognitive approach in natural language processing not only wants to build a machine that appears to understand language, but also *that* actually understands language. In this way, the simulation mode promotes a higher level of anthropomorphism than the performance mode by trying to simulate the operations of human thinking. The objective of the researchers that follow the simulation mode is to accomplish certain human tasks not in the most efficient way (Weizenbaum, 1976), but in the human way. This fact promises to test theories and models about human behavior and cognition through their intelligent agents. The idea is to make computers simulate social, emotional, and cognitive theories to see what inferences can be formulated (Weizenbaum, 1976).

The performance and simulated mode correspond to the two positions within the field distinguished by Searle (1990): strong and weak artificial intelligence. Strong artificial intelligence refers to scientists who confuse external behaviors with inner processes. In other words, they confuse "simulation" with "duplication." Searle explains that a machine may manipulate symbols, but that it doesn't mean that the machine actually understands the symbols or attaches some meaning to them. This position corresponds to the simulation or cognitive mode, also called symbolic approach. These scientists believe "that it is possible to duplicate human intelligence in artificial systems where the brain is seen as a kind of biological machine that can be explained and duplicated in an artificial form" (Duffy, 2003, p. 178). By contrast, weak artificial intelligence refers to scientific theory that recognizes that computers imitate certain mental abilities and do not claim that computers can understand or that are intelligent. Weak "artificial intelligence implies that human intelligence can only be simulated" (Duffy, 2003, p. 178). This description fits better the performance mode due to their interest in emulating just human behaviors instead of the processes that drive them.

The word simulation in the simulation mode or cognitive approach hides that the objective of some of these researchers is not to simulate, but duplicate humans. Simulation, according to the philosopher Baudrillard, "is to feign to have what one hasn't" (Baudrillard, 1983, p. 3). The simulation is to dissimulate an absence, to generate models of the real without referring to the real. In this case, the external behavior of computers simulate that they think or that they have emotions, this simulation implies that they lack of the human inner processes that develop thought. This definition of simulation would apply better to the performance mode, in which robots or agents appear to understand language but do not necessarily understand language. Their simulation of human dialog hides their incapacity for language understanding, according to critics such as Searle (1990). By contrast, in the simulation mode, researchers try to generate a model of human-like communication with reference to real human communication: machines appear to behave human-like because their inner processes are based on humans' mental processes. Perhaps the word simulation is used to refer to a deeper absence: machines seem to understand the meaning of a word, but do they know what it means or even that it means (Postman, 1993)? The same happens in the simulation of human emotions, as Weizenbaum explains: "even if a computer could simulate feelings of desperation and of love, is the computer then capable of being desperate and of loving? Can the computer then understand desperation and love? … the answer is 'no'" (Weizenbaum, 1976, p. 200).

Therefore, the main differences between the weak and strong artificial intelligence are their objectives and how they regard their accomplishments. Weak artificial intelligent researchers who follow the statistical or performance approach want to show human characteristics in their agents by implementing probabilistic models and do not think that these human-like behaviors are real. On the other hand, strong artificial intelligent researchers who follow the simulation or cognitive approach try to emulate not only the external human-like behavior but the internal processes that produce them, and since they are trying to emulate the processes, then the results are seeing as more real than those of the statistical approach. Although researchers from the cognitive or simulation mode consider this approach more human than the statistical one, our brains operate—not in a conscious way—using statistical algorithms that allow us several mental processes from knowing how to catch a ball, consciously ignoring an estimate of its speed and trajectory, to decision-making operations based on intuitions. In addition, not using the statistical approach is ignoring an advantage of computers, their capacity of simultaneously processing several possible outcomes, and their actual probability. Many researchers, however, use both approaches, called hybrid techniques. Using a hybrid approach achieves the best results regarding efficiency and human-like behavior. As an example, IBM's intelligent agent Watson won the *Jeopardy!* contest when competing against the two best humans players, getting a score higher than the sum of the two human scores (Kurzweil, 2012).

The Emulation of Emotions: Approaches and Motivations

The goal of the statistical or performance mode is to build machines that appear emotional, while those that follow the cognitive or simulation approach not only want to build a machine that appears emotional, but also that *has* emotions.

Two examples, the techno-handshake (Duffy, 2006) and ELIZA (Weizenbaum, 1976), are useful to illustrate how researchers following the statistical or performance mode to emulate human characteristics in their inventions. On the one hand, a technology used in robotics called the "techno-handshake" is used in robots to achieve humans' performance in interpreting the emotions of an individual (Duffy, 2003, 2006). This mechanism interprets humans' mood states by measuring blood flows, heart rates, and stress when a robotic hand has contact with a human hand. This technology belongs to the performance mode since it tries to accomplish the same results that humans achieve in perceiving the emotions of an individual by using their intuition, but it accomplishes it by collecting and interpreting information not used by humans. Weizenbaum's (1976) ELIZA also illustrates the performance mode within artificial intelligence. ELIZA was a computer program capable of having a simple typewritten conversation with humans through rephrasing statements and using nouns to construct questions. To limit the conversation within a manageable range of possibilities, ELIZA was designed to parody a psychiatrist. Weizenbaum's goals were not to imitate the mental processes that support human conversation and emotional responses but to build a program capable of maintaining a simple dialog.

By contrast, researchers who follow the cognitive or simulation mode try to emulate human verbal and nonverbal emotional communication by following models and theories from psychology, anthropology, sociology, and linguistics. While many times their agents' behavior is aided by probability rules, most researchers create learning agents that can improve their performance by training with humans. For instance, the objective of MIT's Humanoids Robotics Group (http://www.ai.mit.edu/projects/humanoid-robotics-group/), which has produced emotional robots such as Kismet, is to recreate the natural way in which infants learn to communicate through experience. In their own words: "Our approach is inspired by the way infants learn to communicate with adults ... the mode of social interaction is that of a caretaker-infant dyad where a human acts as the caretaker for the robot" (http://www.ai.mit.edu/projects/sociable/kismet.html). Thus, although they use hybrid techniques to reproduce emotional gestures and emotions in their robots, researchers from this laboratory try to make robots that learn "more sophisticated communication skills" from their interactions with humans. Using this framework, Kismet has been able to represent six different emotions. Emotions in this robot not only improve the human-machine interaction, but also help it to improve its vision system by expressing negative emotions if the individual is too far and positive emotions when the individual is close and easily visible through its cameras.

Cassell's and colleagues' embodied conversational agents (Cassell, Bickmore, Campbell, & Vilhjamsson, 2001; Finkelstein, Ogan, Vaughn, & Cassell, 2013) also belong to the simulation mode. These software agents are virtual humans that through hand and facial gestures, turn-taking processes, bodily positions, prosody in speech, and nonverbal signs, are able to convey emotions in their interactions with individuals. Cassell applies her linguistic and psychological knowledge about human-to-human conversations to implement emotions and turn-taking signs in her embodied agents. Cassell, Tartaro, Rankin, Oza, and Tse (2005) designed a virtual child called "Sam, the Castlemate." Children can play with Sam to narrate stories that take place in a doll's castle, equipped with several sensor apparatuses through which the software agent can track the position of the toys and the child's hands. Cassell has used some of these empathic agents as therapy for autistic children, where the exchange of emotions is a key element (Tartaro & Cassell, 2008).

Similar to Cassell, one of the researchers interviewed tutors virtual agents using hybrid techniques while trying to imitate children's learning abilities to reproduce communication skills through experience. Her objective is to create a virtual tutor that can help students master certain subjects when the professor is not available. She explained that she uses the cognitive approach because that could make her agents appear as more "human" than those generated using solely the statistical approach. She uses humans' emotions and communication skills as a source of inspiration and thinks that by replicating them she is also contributing to their understanding.

Regardless of the differences between their approaches, researchers who emulate emotions argue that these have many uses in human-computer interaction. Just as in natural language processing, researchers explain that to achieve a successful human-machine interaction, machines should be able to understand not only speech or text, but also nonverbal emotional cues. Researchers who make virtual teachers such as Cassell point out that human students learn better and feel more motivated when software agents show emotional behaviors (Cassell et al., 2005; Breazeal, 2010). In this way, emotions promote empathy between the virtual teacher and the student providing a better educational experience. Users also evaluate the tutoring agent as more believable, natural, persuasive, and trustworthy when it perceives their affective state and gives adequate emotional reactions (Baldassarri & Cerezo, 2012).

Breazeal (2007), founder of MIT's Personal Robots Group, confirms the idea that emotional machines are considered more empathic by users with her "sociable" robots, in which individuals' responses and evaluations toward the robot improve when it shows emotional behavior such as facial gestures. Breazeal (2010) argues that humans react to machines in human ways (i.e., just as we react to other humans). Therefore, it would be beneficial to also have human-like responses from the machines. The emulation of emotions would be necessary to improve this interaction, especially in the case of personal robots such as caregivers or assistants.

The emulation of emotions could also be motivated by the idea of enabling the still technologically impossible multitask agents. At this time, computer agents are useful for very specific problems or domains. A multitask system would need to equilibrate and combine several knowledge bases that could be contradictory in some instances. The creation of an artificial personality and the emulation of emotions could help to build a coherent system of thought that is capable of organizing several kinds of knowledge and ideas, just as multiethnic and colonial individuals are capable of building a unified cultural background through syncretism. This is similar to Minsky's redefinition of emotions as ways of thinking (Minsky, 2006). By adding emotions to machines, researchers would be adding more flexibility in the programming since they could follow more paths to solve a problem.

However, researchers have not fully explained the usefulness of imitating some emotions and gestures. One researcher from MIT's Humanoid Robotics Group published that his goal was to "create a foundation for human experiences that are missing from the robotic array such as tiredness, fatigue, and soreness" (http://www.ai.mit.edu/projects/humanoid-robotics-group/people.html). Clearly, these are necessary to make sociable robots, but the question of why researchers want to make them social remains unanswered. Some suggest that the emulation of sociability could be motivated by the objective of substituting a human by a machine in a human-to-human interaction (Duffy, 2003). The objective of emulating humans' sociability is, in the end, a desire to interact with the machine as if it were another human.

Finally, researchers also justify their attempts to replicate human beings by explaining their use of nature as a source of inspiration. The idea of using nature as a source of inspiration is present in many discourses of the cognitive approach. However, trying to replicate nature sometimes obstructs the development of useful and efficient machines: "the most ineffective kind of machine is the realistic mechanical imitation of a man or another animal"

(Mumford, 1934, p. 33). Early inventors of the airplane were inspired by the flight of the birds, but their constructions did not accomplish the desired results until they gave up their idea of imitating birds' wings in their full intricacy. In artificial intelligence, for instance, it is not clear why some researchers invest huge computational resources to construct walking robots for spaces in which the use of wheels could have been more effective and less expensive. This shows how the simulation mode is not interested in results accomplished through nonhuman processes, even if following nonhuman processes is not the most optimal solution to accomplish such results, like applying statistics to calculate probabilities. Hence, the cognitive approach pursues a higher level of anthropomorphism than the statistical or performance mode. Many inventions based on the cognitive approach lead us to think that the emulation of human attributes in artificial agents is not driven by efficient purposes, but by the desire to extend the limits of technological power and human knowledge through human reproduction through artificial ways. In this way, is the creation of anthropomorphic technologies a mean or an end in itself? If it were just a mean, what could be the purpose of researchers who follow the cognitive approach of emulating not just humans' performance but the mental processes behind their behaviors? These inventions could be driven by "virtuosity values" (Pacey, 1983), which pursue technological enterprises not for utility or economic benefits, but to conquer nature and to demonstrate one's technological capabilities. By imitating humans, artificial intelligence scientists try to conquer nature by being able to emulate human beings' attributes, specially their mind.

ANTHROPOMORPHISM IN ARTIFICIAL INTELLIGENCE

Anthropomorphism has been present in artificial intelligence since its beginnings. The early goal of this branch of computer science was to create a machine whose behavior could be equivalent to humans' (Weizenbaum, 1976). Kurzweil (1999) describes the field of artificial intelligence as the art of creating machines capable of performing actions that are considered intelligent when performed by a human. This objective continues to guide artificial intelligence research as evidenced by Ishiguro and Nishio's (2007) goals: to construct a robot indistinguishable from a human in its appearance, movements, perception, and conversation. In this way, there are three levels of anthropomorphism within the design and construction phase of artificial intelligent agents; from the lowest to the highest level of anthropomorphism: emulation of humans' physical appearance, imitation of humans' external behavior (performance or statistical mode), and simulation of humans' mental processes (cognitive or simulation mode).

The emulation of emotions, once considered "program-resistant" functions (Newell, 1982), promote a higher level of anthropomorphism within the field of artificial intelligence, where every day another characteristic considered only human is emulated or modeled in the machine. The redefinition of emotions as information patterns and cognitive percepts is another step in reducing the conceptual differences between the human and the machine. Since emotions are data, they are replicable, which falls in the larger trend within this and other disciplines of considering the human as a replicable entity.

Researchers who follow this anthropomorphism may confuse the possibility of imitating certain human attributes, such as thought or emotions, with its actual understanding. In other words, strong proponents of machine anthropomorphism believe that if they can imitate something is because they understand it, and vice versa (Weizenbaum, 1976). Simulation, in this case, is taken as duplication. Humans' replicability challenges popular religious beliefs like the existence of a soul or spirit, life after death, human freedom and, ultimately, the existence of a God. Minsky (Boston Globe, 2006) completely rejects the belief that humans possess a soul: "we prefer the idea that inside ourselves is some sort of spirit ... that ... feels and thinks for us ... your laptop computer has billions of parts, and it would be ridiculous to attribute all its abilities to some spirit inside its battery." As Minsky, some of the researchers interviewed rejected the possibility of a human soul as can be seen from some of their remarks: "I don't believe in human souls"; "we live in a deterministic world"; "maybe there is a unique thing, we are flesh and they are silicon"; "we are preprogrammed just as the computers"; "if you are stubborn enough, you can simulate biology, reproduction, intimacy." These phrases show how these researchers think that the human being could be replicable. In the future, when technological developments and scientific knowledge about the brain improve, it will be possible to emulate all human attributes. If humans do not have a soul or freedom, probably all their attributes can be emulated. The notion of replicability reduces all human attributes to empirical phenomena, things that can be perceived through our senses, and rejects the possibility of all metaphysic objects. The belief in humans as totally replicable entities implies determinism. In a determined world, there is no place for God or for human freedom. If human behavior and mental processes are fully computable, they are also predictable. Just as "intelligent" machines that start with a set of rules and then adjust their actual performance based on recorded information from past experiences, humans become prisoners of their brain design that is being shaped by the input received from the environment through their sensory organs in a feedback loop form. This implies that the only thing that separates scientists from replicating humans is their lack of knowledge of how the brain works and the technologies to apply this knowledge. In a world without God, anthropomorphism could be the path to fulfill humans' ancestral dreams "for God-like creativity" (Feenberg, 1991, p. 109) "to reproduce nature by artificial means," (Zdenek, 2003, p. 341) and to obtain the power to artificially recreate life. This driven trend within artificial intelligence promises to supersede God's creatures, made in his own image, by creating something not similar, but rather equivalent to humans.

The emulation of emotions can provide the necessary elements to pass the Turing test since the machine would be able to answer truthfully about emotions if they are duplicated. In addition, the emulation of emotions would provide the bridge for the final merging of the human and the machine in the Singularity proposed by Kurzweil (2005). He predicts that humans will become augmented-humans in the following years by introducing technology into their bodies to extend some of their capacities and finally, to avoid death, will be able to download their information patterns to a machine, which would mean a total anthropomorphism of the machine and a total machine-morphism of the human.

REFERENCES

Baldassarri, S., & Cerezo, E. (2012). Maxine: Embodied conversational agents for multimodal affective communication. *Computer graphics*. http://www.intechopen.com/books/computer-graphics/maxine-embodied-conversational-agents-for-multimodal-affective-communication.

Baudrillard, J. (1983). *Simulations*. New York: Semiotext, Inc.

Boston Globe. (2006). *Minsky talks about life, love in the age of artificial intelligence*. www.boston.com/news/globe/health_science/articles/2006/12/04/minsky_talks_about_life_love_in_the_age_of_artificial_intelligence/.

Breazeal, C. (2007). Sociable robots. *Journal of the Robotics Society of Japan*, 24(5), 591–593.

Breazeal, C. (2010) The rise of personal robots. *TED talks*. http://www.youtube.com/watch?v=eAnHjuTQF3M.

Cassell, J., Bickmore, T., Campbell, H., & Vilhjamsson, H. Y. (2001). More than just a pretty face: Conversational protocols and the affordances of embodiment. *Knowledge-Based Systems*, 14, 55–64.

Cassell, J., Tartaro, A., Rankin, Y., Oza, V., & Tse, C. (2005). *Virtual peers for literacy learning*. Retrieved Dec. 2007 from www.inagreendaze.com/Papers/Cassell.EdTech.Feb6.pdf.

Damasio, A. (2005). *Descartes' error: Emotion, reason, and the human brain*. New York: Penguin.

Duffy, B. R. (2003). Anthropomorphism and the social robot. *Robotics and Autonomous Systems*, 42, 177–190.

Duffy, B. R. (2006). Fundamental issues on social robotics. *IRIE International Review of Information Ethics, 6*, 31–36.
Feenberg, A. (1991). *Critical theory of technology*. Oxford: Oxford University Press.
Finkelstein, S., Ogan, A., Vaughn, C., & Cassell, J. (2013). Alex: A virtual peer that identifies student dialect. In *Proceedings of workshop on culturally-aware technology enhanced learning, Paphos, Cyprus*.
Goya-Martínez, M. (2008) *Anthropomorphism in artificial intelligence: Human emulation, human metaphors, and the redefinition of human*. Unpublished manuscript.
Hayles, N. K. (2008). *How we became posthuman: Virtual bodies in cybernetics, literature, and informatics*. Chicago, IL: University of Chicago Press.
Human Brain Project (2015). https://www.humanbrainproject.eu/.
Ishiguro, H., & Nishio, S. (2007). Building artificial humans to understand humans. *Journal of Artificial Organs, 10*, 133–142.
Kurzweil, R. (1999). *The age of spiritual machines: When computers exceeded human intelligence*. New York: Penguin.
Kurzweil, R. (2005). *The singularity is near: When humans transcend biology*. New York: Penguin.
Kurzweil, R. (2012). *How to create a mind: The secret of human thought revealed*. New York: Penguin.
McLuhan, M., & Fiore, Q. (1967). *The medium is the massage*. New York: Bantam Books.
Minsky, M. (1988). *Society of mind*. New York: Simon and Schuster.
Minsky, M. (2006). *The emotion machine: Common sense thinking, artificial intelligence, and the future of the human mind*. New York: Simon & Schuster.
Moravec, H. (1988). *Mind children: The future of robot and human intelligence*. Cambridge: Harvard University press.
Mumford, L. (1934). *Technics and civilization*. New York: Peter Smith.
Newell, A. (1982). *Intellectual issues in the history of artificial intelligence*. Pittsburgh, PA: Research Showcase @ CMU.
Pacey, A. (1983). *The culture of technology*. Cambridge: The MIT Press.
Postman, N. (1993). *Technopoly: The surrender of culture to technology*. New York: Knopf.
Randell, B. (1982). From analytical engine to electronic digital computer: The contributions of Ludgate, Torres, and Bush. *Annals of the History of Computing, 4*(4), 327–341.
Schelling, T. (1960). *The strategy of conflict*. Cambridge, MA: Harvard University Press.
Searle, J. (1990, January). Is the brain's mind a computer program? *Scientific American, 26*.
Shelley, M. (1818). *Frankenstein, or the modern prometheus*. London: Lackington.
Tartaro, A., & Cassell, J. (2008). Playing with virtual peers: Bootstrapping contingent discourse in children with autism. In *Proceedings of international conference of the learning sciences (ICLS), June 24–28, Utrecht, Netherlands*.
Turing, A. (1999). As we may think. In P. A. Mayer (Ed.), *Computer media and communication, a reader* (pp. 23–36). Oxford: Oxford University Press. Original edition 1950, Chap. 1.
Weizenbaum, J. (1976). *Computer power and human reason*. San Francisco, CA: Freeman and Co.
Wiener, N. (1965). *Cybernetics or control and communication in the animal and the machine*. Cambridge: The MIT Press.
Zdenek, S. (2003). Artificial intelligence as a discursive practice: The case of embodied software agent systems. *AI & SOCIETY, 17*, 340–363.

CHAPTER 9

Through Google-Colored Glass(es): Design, Emotion, Class, and Wearables as Commodity and Control

Safiya Umoja Noble[a], Sarah T. Roberts[b]
[a]Department of Information Studies, Graduate School of Education & Information Studies, University of California, Los Angeles, CA, USA
[b]Faculty of Information and Media Studies, The University of Western Ontario, London, ON, Canada

INTRODUCTION: IN THE GOOGLE GAZE

We were trying to learn about social issues around Glass ... we ended up sending signals that it was a finished product, like putting it on a (fashion) runway ... I wish we had done differently.

(Astro Teller, Head of Google X)[1]

In April 2104, the research firm Toluna reported that 72% of the public hated Google Glass, a wearable personal information assistant technology, due to privacy concerns based on its capacity for surveillance.[2] Multiple reports of Google Glass wearers being attacked for wearing the technology in public have made headlines across the United States since Google first launched its exploratory beta to select users in spring, 2013. But what is Google Glass, and why does it engender such emotion among those who both love and hate the technology? In this paper, we discuss the implications of wearable technologies like Google Glass that function as a tool for occupying, commodifying, and profiting from the biological, psychological, and emotional data of its wearers and, critically, from those who fall within its gaze. We also argue that Google Glass has a fundamental design flaw that privileges an imaginary of Whiteness and unbridled exploration and intrusion into the physical and emotional space of others. The dominant narrative of Google Glass, as evinced through its marketing strategy, posits its wearers

---

[1] As reported by Alistair Barr for the *Wall Street Journal*, http://blogs.wsj.com/digits/2015/05/29/googles-moonshot-chief-says-early-version-of-glass-wasnt-ready/.
[2] Matyszczyk (2014, April 8).

as "Explorers," a familiar colonial narrative. This disposition maps onto the current processes of radical gentrification and displacement in many San Francisco neighborhoods, and signals that power, Whiteness, and class elitism are core values in the Google Glass design imaginary. Glass's recognizable esthetic and outward-facing camera has elicited intense emotional response, particularly when "exploration" has taken place in areas of San Francisco occupied by residents who were finding themselves priced out or evicted from their homes to make way for the techno-elite.

Google Glass is a head-mounted, wearable optical technology that promised users to "take pictures, record what you see hands-free, share what you see live, obtain directions, send messages, and ask whatever is on your mind."[3] With a series of Google and third-party applications, Google Glass promised to be a technology that would allow the wearer to scan images, and people, in the user's line of sight and "Google them," using the massive data power of Google search engine, to provide more information to the wearer about objects and people captured in the Google Glass gaze. Google marketed it as tool of freedom, with a powerful video depicting a series of experiences that can be captured "hands-free" to foster a greater sense of participation and power over one's informational and geospatial environment than one can experience holding a smartphone (Figures 9.1 and 9.2).

Utopian discourses of freedom through technology are not new. Critical geography scholars have written extensively about how hierarchies of power are reproduced and enacted through digital technologies (Crampton & Krygier, 2006; Goss, 1995; Harvey, Kwan, & Pavlovskaya, 2005; Noble, 2011; Schuurman, 2000), and they point to the ways in which the informationalization and digitization of everyday life heightens control and surveillance, but also establishes digital enclosures (Andrejevic, 2007) with power-laden boundaries across race, gender, and class. Technological projects are never neutral (Pacey, 1983; Winner, 1986), and Google Glass is not a tool of freedom, despite its marketing discourse.

The most controversial aspect of Google Glass has been its outward-facing camera, allowing users to record things within their field of vision, which the camera then follows. Concerns have ranged from the potential theft of ATM passcodes by passersby wearing Glass, to the illegal infringements of Glass wearers recording copyrighted material in movie theaters, or the potential for cheating in Las Vegas casinos, which have banned wearers. Video resulting from all of these potential uses of the technology could easily

---

[3] See: "What it does," *GLASS*, http://perma.cc/Q5SH-BPWP.

Through Google-colored glass(es)   189

Figure 9.1 The Google Glass interface, from the user perspective, as depicted in a Google Glass promotional video. Source: *Why brands are already looking at Google Glass, and why Apple should be worried*. AdWeek, 2013, February 20. http://www.adweek.com/news/technology/why-brands-are-already-looking-google-glass-and-why-apple-should-be-worried-147435.

Figure 9.2 Screenshot of a frame of "How it Feels [through Google Glass]" promotional video. Source: *Why brands are already looking at Google Glass, and why Apple should be worried*. AdWeek, 2013, February 20. http://www.adweek.com/news/technology/why-brands-are-already-looking-google-glass-and-why-apple-should-be-worried-147435.

be uploaded real-time or streamed to the Internet to be circulated, stored, and owned by Glass wearers, all without the consent of anyone within the gaze of Google Glass. Consistently, critics and news reports have largely leveled critiques at the product as an important site of the fight to resist to hypersurveillance and privacy concerns.[4]

UNEXAMINED OCCUPATION: SAN FRANCISCO AND THE CRYSTALLIZATION OF GLASS AND CLASS RAGE

Google Glass arrived at a time when Google, headquartered in Mountain View, California, had already incited people feeling the economic and emotional ramifications of the changing landscape of the San Francisco Bay area as the latest tech boom has pushed people out of affordable housing *via* evictions. Landlords across California, but particularly in San Francisco, have taken great advantage of the Ellis Act[5] during the most recent techno-boom in housing, allowing them to clear all tenants out of their multiple unit dwellings, often rent-controlled, by claiming that landlord is "going out of business." This strategy is typically used to convert apartment building into high-priced condominiums. In other cases, landlords cleared out buildings multiple times under the guise of "going out of business" to perpetually increase rents. These practices have made new and existing housing unaffordable for those residents not employed in high-paying sectors such the tech industries or finance. Indeed, *Newsweek* reported, "In 2005 … after new technology companies like Google began attracting thousands of high-paid employees to the Bay, the number of Ellis evictions tripled. In 2013, Ellis evictions grew 175% from the year before" (Kloc, 2014).

Google, in particular, has received special ire not only because of perceptions of its presence being responsible for Ellis evictions, but also due to its ubiquitous fleet of luxury buses and their impact on life in the Bay Area. These coaches move San Franciscans who labor at its Mountain View corporate headquarters to and fro every day in private Wi-Fi-enabled, air-conditioned comfort. According to the Master's thesis of City Planner Alexandra Goldman, rents at apartments near Google Bus stops rose 20% during a period where the average rent increase was 5%. Included in her

---

[4] See: Will Oremus, "'Don't be creepy': Google Glass won't allow face recognition," *SLATE* (2013, June 3), http://perma.cc/NE2E-QW92.
[5] A detailed description and maps of San Francisco have been collected by anti-eviction activists at http://www.antievictionmappingproject.net/ellis.html.

evidence were Craigslist ads denoting the presence of the Google Bus stops as a perk for apartments located nearby. Apple, eBay, Twitter, Electronic Arts, Facebook, and Yahoo!, among others, also run private bus lines, which have been protested by local residents.[6]

Indeed, Google and many high-tech companies in Silicon Valley have directly contributed to gentrification in previously multiracial neighborhoods in the San Francisco Bay Area. A challenge of gentrification is that it brings new wine bars, vape shops, restaurants, and boutiques to areas that might previously have not had such high-end amenities. But, gentrification has always been a narrative of "improvement" that draws upon a mystique of transforming previously "uninhabitable" places into spaces that can be occupied. Like colonial projects of the past that seek an expansion into new territories and locales, despite there already being inhabitants of those spaces, gentrification is hailed as offering an improvement or reinvestment into neighborhoods that, when occupied by low-income people of color, were seen as not valuable or necessarily productive. As San Francisco is overrun with tech workers, the majority of whom are White and Asian men with lucrative salaries and stock options (Sullivan, 2014), the city is experiencing "hypergentrification," a term coined by policy analysts who watched the millionaires produced at local tech giant, Twitter, buy up all the middle-income housing, displacing urban residents (Figure 9.3).[7]

While the private coach services do offer reduced greenhouse gas emissions versus all 4500 riders taking private vehicles, they have also monopolized local public transit bus stops in the MUNI system, initially without remuneration to the city (Goldman, 2013), and are part of a series of other notorious soft benefits at Google that appear egregious to those who do not work in similar tech industry settings.[8] In short, the Google Bus and the Google employees it ferries back and forth to Silicon Valley have become highly visible symbols of class division in San Francisco. In response, locals

---

[6] See: "Protesters block Apple, Google buses in San Francisco area" by Alexei Oreskovic at http://www.reuters.com/article/2013/12/21/us-techbus-protest-sanfrancisco-idUSBRE9BJ1BC20131221.

[7] See: "Twitter will cause so much gentrification, they invented a new word" by Nitasha Tiku at http://valleywag.gawker.com/twitter-will-cause-so-much-gentrification-they-invente-1447346147.

[8] Of course, these perks are designed, in large part, to keep Google employees working as long as possible; the bus transport with its Wi-Fi turns commuter time into work time. See more on Google perks here: http://www.businessinsider.com/google-employee-favorite-perks-2013-3?op=1.

Figure 9.3 San Francisco activists protest private corporate buses' use of local public transportation infrastructure, 2013 December. Source: *http://www.flickr.com/photos/cjmartin/11295749384—Creative Commons license.*

have gone so far as to block buses from moving or stopping for pickups, and smashing bus windows.[9]

These tensions and negative emotions reflect the growing distrust between the discourses of Google as a company for everyday people, providing seemingly "free" services that improve the quality of our lives while asking for nothing in return. Jay McGregor, writing for Forbes.com said, "Google has suffered an image problem since it was accused of providing a backdoor to the National Security Administration (NSA) in Snowden's documents. Since then it has gone on a campaign to slowly regain the public's trust in carefully managed bursts of 'honesty.'"[10]

---

[9] See: http://bits.blogs.nytimes.com/2013/12/20/google-bus-vandalized-duringprotest/?_php=true&_type=blogs&_r=0.

[10] See: "Good guy Google goes on love bombing initiative to win back public trust" by Jay McGregor at http://www.forbes.com/sites/jaymcgregor/2014/06/05/good-guy-google-goes-on-love-bombing-initiative-to-win-back-public-trust/.

The real-life contradictions, like selling the public out to the NSA, or gentrifying low-income neighborhoods of color, are where the imaginary of Google's utopian benevolence are made most apparent. Likewise, Google Glass occupies a similarly visible and elitist cultural space: made available only through a limited beta program afforded to the wealthy and connected. While other Google products' functions of data mining and usage tracking are undertaken on an individual user basis (e.g., in one's own Gmail account), Google Glass is the visible manifestation of Google's tracking turned outward onto others, just as Google buses are the outward demonstration of neighborhood occupation and urban colonization. In a rapidly gentrifying and increasingly economically stratified San Francisco, people have reacted with anger and insecurity when finding themselves the unwilling targets of Google Glass's gaze. In this context, the Glass users have been received in San Francisco and cities like it not only as Glassholes, but as Classholes, too.

GOOGLE GLASS'S PANOPTIC GAZE: SURVEILLANCE AND EMOTION

Michel Foucault's canonical Discipline and Punish (1977), with special emphasis on the notion of Panopticism, provides both a historical basis and an ongoing metaphor for understanding the development and normalization of surveillance, control, and the birth of a prison class; indeed, many digital media scholars have employed this metaphor to describe the nature of the contemporary surveillance state. In his treatment of discipline and punishment within the institution of the prison, Foucault provides evidence for the ways in which control and power were, in that context, mechanized, routinized, and formulated in terms of technological and organizational processes that could be reproduced (architecturally), extended (psychologically, within the prisoner's psyche), and therefore made ubiquitous. The logical extent of this paradigm has been the dual development of the perpetual criminal/prison class, on the one hand, and the process of normalization of surveillance and permanent shifting of the power relationship in favor of the watcher, on the other.

In contemporary society, the latter has extended beyond the physical confines of the prison such that individuals frequently engage in surveillance—even self-surveillance—without even being aware of it. This process is now almost always enabled by digital technology and extends beyond surveillance cameras in places like shopping malls, street corners, and other public and private spaces alike (Crawford, 1992; Judd, 1995; Koskela, 2000; Shields, 1989) to include panoptic invasions of privacy through documentation (photographic and audio) of people in what might have been previously considered private experiences conducted in public spaces. The ubiquity of making the previously mundane and anonymous experience of living in public spaces is increasingly being diminished through multiple modes of surveillance, facilitated by Internet-based reproduction and dissemination. We see evidence of this in the rise of new social phenomena that include "selfies," photography, viral videos (with and without the permission of those digitally recorded), and even in calls for body cameras to be put to use by law enforcement after extrajudicial killings of unarmed men, women, and children. Add to this list the constant self-surveillance and digital production of monitoring of one's emotional and physical health through new technological devices like the Fitbit, the Apple Watch and its Health app, and myriad other similar devices that record, track, analyze, and share what was once considered personal data, or what was once not considered at all.

Foucault (1977) begins his discussion by describing the architecture of the physical Panopticon as envisioned by its creator, Jeremy Bentham. Bentham's intent was to create a solution allowing for the surveillance and monitoring prisoners with a modicum of staffing and at reduced costs. But there were further benefits to the physical layout he proposed (center tower and cells located on the periphery of a circular building surrounding that tower, disallowing the occupant from seeing anything but the watchtower located in the middle of the space); the individual prisoner, in a Panopticon, "… is seen, but he does not see; he is the object of information, never a subject in communication" (p. 200). Further, Bentham's Panopticon was intended to create a particular power structure, in which power (of the watcher/guard) would be both "… visible and unverifiable" (p. 201), creating in the prisoner the seemingly paradoxical, and certainly unnerving, sense of both being constantly watched and yet never being able to verify the watching. Because of this lack of verifiability, the prisoner would therefore be forced to assume that he was being watched at all times, and behave accordingly, becoming a party to his own surveillance and control in the process. Meanwhile, the role of the watcher, when filled, would be afforded both anonymity and the practical and efficient interchangeability of workers. This feature would result in the "… automatiz[ing] and disindividualiz[ing of] power" (p. 202) from those meting it out.

This technological intervention (Foucault describes it as a "machine" [p. 202]) disturbed the bidirectional relationship between the seer and the seen, and created a sense of constant, yet unverifiable, surveillance. At the same time, the eighteenth century gave rise to an entire prison class of "… people who were believed to be criminal and seditious as a whole," (p. 275) coming primarily from the lower social classes (p. 275). There was therefore a symbiotic and circular relationship created among the less privileged social classes of imprisonment, surveillance, and control that was both borne of and reinforced their position in the lower social strata, and that encouraged, if not mandated, an acquiescence to a paradigm of being watched and of, in turn, self-regulating under the ubiquitous potentiality of surveillance.

While the physical layout of the Panopticon prison has fallen out of favor in contemporary society, its notions and influence endure; although the architecture of the prison may no longer match Benthem's vision, the notions of control and power exerted on individuals through a constant potential for and expectation of surveillance is still key to behavioral management in prisons, as is isolation (a classic example would be the Special Housing Unit of California's notorious "SuperMax" prison, Pelican Bay, or the 43-year solitary confinement of former Black Panther Albert Woodfox in Angola, Louisiana). This practice of self-policing under a paradigm of potential and anonymized surveillance has expanded and spread into everyday life (e.g., the urban closed-caption TV cameras recording the movement and actions of people in the course of their daily lives), and is emblematized through projects like Google Glass, which is rarely framed as a panoptic project, but certainly engenders a similar class distinction as wearers become the watchers, and those without Glass become the watched.

Further, the exponential growth of a prison class made up of persons marked "by a series of brandings" (p. 272), for and by institutionalization and confinement by virtue of their social, racial, or ethnic class or status, lack of educational opportunity, and so on, continues unabated as notions of what can and should be criminalized grow wider and penalties for infractions increase. Panoptic power attempts to reduce emotional expressions and is a racialized, class, and gendered project. Those who are more likely imprisoned and victimized by the gaze of surveillance technologies are poor, often people of color, and women, and never is this more evident than in the ways that digital observations of "the other" are deployed in contemporary United States. Here we see these practices on full display with ever more people passing through the machine of the prison, more and more people therefore acculturated to its surveillance and control, and increasingly fewer people left untouched or outside of the regime of the prison and surveillance culture, able to see, respond to, and resist the demands of the anonymous gaze of Panoptic power.

In the case of Google Glass and facial recognition software, Google executives often argue in the media that these forms of surveillance are technically possible but would not be a feature of the product unless privacy concerns could be protected. But facial recognition technologies, for example, have previously been implemented by law enforcement agencies[11] for use in centralized neighborhood surveillance systems (Sarpu, 2014), such as cameras installed on telephone and streetlight poles, and these tend to be located in hypermarginalized, poor communities that are predominantly comprised of non-White people, in programs known by names such as "broken-window" policing (Wilson & Kelling, 1982). Sarpu (2014) documents how government agencies have also relied upon facial recognition and video surveillance systems in border patrols and crossings like the United States Visitor and Immigrant Status Indicator Technology (US-VISIT) program, or in "voter fraud" projects to screen people out of voter participation. Each of these deployments is primarily targeted by police on Black and Latino bodies in their engagements with the state through border crossing or voting.

Although Google has stated that it would reject facial recognition apps, Sarpu has brought attention to the third party applications, or "Glassware," that can be installed on Google Glass devices, rendering the company's declaration that it will not load facial recognition technology in the standard out-of-the-box product[12] mostly irrelevant. This then, is the context within which we analyze Google Glass, and the attendant responses by various members of the public and tech community to reject the project, although these rejections have focused less on racial and class dimensions of surveillance. Yet undeniably, Google Glass has been deployed as a project in service of power elites (those less affected by hypersurveillance), yet the consequences of normalizing the loss of privacy *via* heightened surveillance, the differing consequences of surveillance on people dependent on race, socioeconomic status, and other identity-related factors, and the inability for all to equally access technologies like Glass lead to a disregard for the types of harm that are disproportionately impacting poor people, people of color, and women. These issues should be of paramount concern as the product is being refashioned for greater public acceptance.

---

[11] See: Kevin Bonsor and Ryan Johnson, "How facial recognition systems work," *Howstuffworks*, http://perma.cc/RB53-YF28.
[12] See: Eric Larson, "Google glass won't have facial recognition apps yet," *Mashable* (2013 June 1), http://perma.cc/QA65-Y43D.

EMOTION AND RESISTANCE: THE EMERGENCE OF THE "GLASSHOLE" AND PUBLIC PUSHBACK TO GOOGLE GLASS

A crucial concern about wearers of Google Glass has been the intrusion upon others who do not want to be recorded without their knowledge and permission by its wearers. In many cases, the public has responded, popularly terming those who show up in public spaces wearing the technology as "assholes" who wear Glass, or "Glassholes." In some cases, the hostility is based on the arguments made by privacy advocates that Glass applications, allow their users to scan, tag, and link people without their knowledge to anything on the Internet (Miller, 2013; Sarpu, 2014), or even record and disseminate private conversations over the web. Other reasons for backlash also include people's affective response to being targeted by the gaze of an elitist technology not available to all. Whatever the motivation, resistance to Glass has led to various public displays of backlash that have included the banning of the technology from businesses, including restaurants and movie theaters. Artist Julian Oliver created the "Glasshole Free" logo and software[13] as a free download that can run on a mini-computer like the Raspberry Pi to help establishments communicate disapproval and even emit an alarm to alert the presence of Google Glass nearby (Greenberg, 2014). In addition, the "Google Glass is Banned from these Premises" campaign developed a free script, Aircrack-NG, which can be used to kick Glass wearers off the Wi-Fi network (Figure 9.4).

Figure 9.4 *"Don't be a Glasshole" project. Source: Julian Oliver.*

---

[13] See: http://julianoliver.com/output/log_2014-05-30_20-52.

Google Glass violence has erupted in Seattle, San Francisco, and Paris, with wearers subjected to physical beatings, such as the case of the Professor Steve Mann, the "father of Wearable Computing"[14] from the University of Toronto, who was reportedly physically accosted and thrown out of a McDonald's in Paris for wearing a Google Glass-like wearable recording device of his own making while on vacation.[15] Mann, a legendary computer engineer, has been working on an augmented reality Digital Eye Glass through his wearable projects that predate Google Glass by nearly 30 years. Mann's work at the MIT Media Lab, and the founding of the MIT Wearable Computing Project, has been focused on developing what he has coined, "sousveillance,"[16] (Mann, Nolan, & Wellman, 2003), which is making video surveillance a part of the everyday human experience in which all can participate. Mann has intended his sousveillance projects as a mechanism to democratize surveillance, making it accessible to those without the authority to survey from on high. Yet projects like Google Glass, and Mann's digital eyewear and devices are both far from embraced in the public imaginary, and likely with good reason: these wearables, with their heightened capacity for external surveillance of others, provoke a sense of ambiguity about their use and purpose, and a discomfort with the lack of control that individuals have over those aiming their always-on lenses at them.

Sarpu (2014) has thoroughly detailed the history of privacy protections in the United States through key legal decisions that include a right to anonymity[17] as a fundamental aspect of the right to political dissent. Beyond the import of privacy as a protection and the ways in which, at one time for example, Kodak's instant photography was legally ruled to be an invasion of privacy,[18] is the previous legal tendency to protect the public from undue intrusion in the private sphere of life. Ultimately, legal scholars have argued that individuals in the public have "the right to be let alone" and protected from technological encroachment on their private lives, with the exception

---

[14] See: http://wearcam.org/biowaw.htm.
[15] See: http://www.forbes.com/sites/andygreenberg/2012/07/17/cyborg-discrimination-scientist-says-mcdonalds-staff-tried-to-pull-off-his-google-glass-like-eyepiece-then-threw-him-out/.
[16] See: http://en.wikipedia.org/wiki/Sousveillance.
[17] See: Daniel J. Solove, *The Future of Reputation: Gossip, Rumor, and Privacy on the Internet* (Yale University Press, London, 2007), pp. 139–140.
[18] See: Samuel D. Warren and Louis D. Brandeis, "The right to privacy," 4 HARV. L. REV. 193 (1890).

of those "who ... have renounced the right to live their lives screened from public observation,"[19] (Sarpu, 2014), which generally means, in this contemporary moment, celebrities, politicians, and others who choose a public life—a public life of their choosing, or those who engage in private activity in public spaces. Invasions of privacy are also not protected for those engaging in newsworthy or notable activity that to preclude it from the public would be against the public interest, which Sarpu notes has consistently been upheld by the Supreme Court. Over time, the public has lost considerable ground in the right to privacy in public spaces, and this is an essential feature of the Google paradigm, as often quoted by Google executives who consistently argue that if one has nothing to hide, there should be no concerns about being publicly surveyed.

The "nothing to hide" position, however, is flawed on multiple levels, as noted legal scholar Daniel Solove (2007) has effectively argued. Google's CEO, Eric Schmidt, is famous for saying, "If you have something that you don't want anyone to know, maybe you shouldn't be doing it in the first place"[20] as a way of undermining the many individuals who have levied complaints about how they are misrepresented by Google. Noble (2013a) has also brought attention to the ways in which the public, particularly marginalized groups like women and girls of color, are unable to effect any change in the ways in which they are misrepresented and sexually objectified in Google's search engine results. This is due, in part, because the public is generally unaware of the ways in which search results appear in Google, and they have little understanding of the multiple layers of bias, curation, and content moderation that are deployed to keep some kinds of information open and available in digital media platforms. It is also not widely understood that content moderation practices often work in the service of large corporations, with little regard for the impact on individuals, as Roberts (2015, 2016) has extensively researched. Google has effectively worked to convince the public that issues like invasion of privacy and inaccurate information in its search engine results would not be a problem for anyone unless they are doing something that they would need to hide. In fact, with few exceptions like the public backlash against Google Glass, Google is erroneously seen as a public good (Vaidhyanathan, 2011). Solove has often disputed Google executives' position by denaturalizing the notion that privacy should be of no

---

[19] See footnote 18.
[20] See: http://www.onthemedia.org/story/260644-if-youve-got-nothing-hide-youve-got-nothing-fear/.

concern to those who do no wrong. Solove rhetorically engages critics who argue that Google and the proliferation of information about us on the Internet is not invading our privacy with a series of questions that underscore the importance of privacy:

> So my response to the "If you have nothing to hide ..." argument is simply, I don't need to justify my position. You need to justify yours. Come back with a warrant. I don't have anything to hide. But I don't have anything I feel like showing you, either. If you have nothing to hide, then you don't have a life. Show me yours and I'll show you mine. It's not about having anything to hide, it's about things not being anyone else's business. Bottom line, Joe Stalin would [have] loved it. Why should anyone have to say more?
>
> *(Solove, 2011)*

Solove's key points focus on protecting the rights of the public and individuals from a reframing of privacy that desensitizes us from recognizing how important privacy is to our quality of life, and to our ability to ensure a democracy free from governmental surveillance.

Surveillance, of course, is largely misunderstood when it comes to Google and its role as privatized information gatherer for the United States government. Nafeez Ahmed recently wrote a two-part investigative journalism piece for medium.com in which he details the ways in which Google emerged as a project funded by the Central Intelligence Agency.[21] The role of Google as an arm of the state in privatizing and monitoring public information through the data mining of users in its products like Gmail, Maps, and search history, is problematic due to its embedded relationship with the NSA. Ahmed is one of many raising legitimate concerns for the public about Google's private and secret relationship with the NSA, which is only on the periphery of public emotion and outcry against the corporation. Ahmed opens the second part of his exposé on Google with the following:

> Mass surveillance is about control. Its promulgators may well claim, and even believe, that it is about control for the greater good, a control that is needed to keep a cap on disorder, to be fully vigilant to the next threat. But in a context of rampant political corruption, widening economic inequalities, and escalating resource stress due to climate change and energy volatility, mass surveillance can become a tool of power to merely perpetuate itself, at the public's expense
>
> *(Ahmed, 2015)*

---

[21] See: "Why Google made the NSA: Inside the secret network behind mass surveillance, endless war, and Skynet" at https://medium.com/insurge-intelligence/why-google-made-the-nsa-2a80584c9c1.

What is most crucial about Ahmed's detailing of the corporate sector's increasing influence on the normalization of mass surveillance, which includes the socialization of the public to engage in panoptic behavior, is that it is the only constitutional and legal way to circumvent laws that protect the public from illegal surveillance. On the heels of evidence provided by former government contractor Edward Snowden about the unparalleled mass surveillance program of the United States government against its own citizens, public outrage is far less visible, particularly in comparison to the public reaction to Google Glass. Using techniques such as privatization, contracting, and deregulation, corporate contractors have taken on the process of creating, managing, storing, and disseminating (or hiding, in some cases) vast amounts of information. Indeed, the recent "Top Secret" report in the *Washington Post*[22] reveals that there are now over 2000 private firms engaged in data analysis for the purposes of national security alone, with little, if any, public redress available to learn more or understand what these firms do.

The foreshadowing of these threats to democracy was articulated by Herbert Schiller in his 1996 work, *Information Inequalities*, which focused on the great shift in power and control from state to private actors, resulting in a massive consolidation of power in the corporate sector, particularly over the control and dissemination of communication and information. Nearly 20 years old, this essay draws out the peculiarity of this new power structure and highlights the disturbing characteristics of that shift, including the crystallization of the already-underway processes (in the United States and, by extension, abroad, wherever the transnational influence of these companies reaches) such media conglomeration, leaving important informational functions, vital to a vibrant democracy, in the hands of a relatively elite few with considerable neoliberal agendas of their own.

The results of the shift from state to private hands have immense and critically important ramifications, Schiller convincingly argues. One major arena of this transformation occurs in the context of an increase in the technologically facilitated disappearance of some information (such as the case of that at the federal level in the context of changing administrations), and the lack of transparency and accountability under new privatized paradigms where private corporations stand in for the government/state.

Adding to the complexity, corporations are afforded protections as individuals, and work to promote a sense of their marketing and communications

---

[22] See: "Top secret America: *A Washington Post* investigation" at http://projects.washingtonpost.com/top-secret-america/.

and individual self-expression. Attempts at control or censorship of individual expression were once seen as only possible by the state, leaving corporations unencumbered from the burdened perception that they have any power or ability to intrude upon the individual. "Where once there was justified fear of government control and censorship of speech, today there is a new form of censorship, structurally pervasive, grounded in private concentrated control of the media, and generally undetectable in a direct and personal sense" (Schiller, 1996, p. 45). Schiller elaborates on the ways in which, "Corporate speech has become the dominant discourse ... While the corporate voice booms across the land, individual expression, at best, trickles through tiny constricted public circuits. This has allowed the effective right to free speech to be transferred from individuals to billion dollar companies which, in effect, monopolize public communication" (p. 45). Schiller, as the environment in which the national information infrastructure has been eroded, cites privatization, deregulation, and the expansion of market relationships. It is in this context that new digital devices, including technologies of surveillance, are being designed and normalized in the interests of corporate profits over public interests, and the erosion of rights, like that of privacy, are fomented.

We argue that the most impactful and wide-reaching of these intrusive corporate projects include the products and services of Google, whose products are designed and predicated upon the ability to maximize data collection on its users through acculturating technologies like Gmail, the Android platform, and a host of automated and wearable technologies of the future.

WEARABLE CONTROL

> With your permission you give [Google] more information about you, about your friends, and [Google] can improve the quality of searches. [Google doesn't] need you to type at all. We know where you are. We know where you've been. We can more or less know what you're thinking about.
>
> **(Eric Schmidt, former Google CEO)[23]**

In May 2013, the United States congressional members of the Bi-Partisan Privacy Caucus met to explore privacy concerns for the public, particularly with facial recognition software that could be used in Google Glass. The congressional committee issued a letter to Larry Page, CEO of Google, inquiring about the privacy implications of the project, and posing several

---

[23] See: Derek Thompson, Google's CEO: "The laws are written by lobbyists," *The Atlantic* (2010, October 1), http://perma.cc/A3PN-3XFT.

questions to the company, the first of which reflected the general sentiments of the committee:

> In 2010, it was discovered that Google was collecting information across the globe from unencrypted wireless networks.[24] This practice caused multiple investigations into the company along with consumers left perplexed.[25] Google just recently agreed to pay $7 million to settle charges with 38 states for the collection of data from unprotected Wi-Fi networks without permission.[26] Google also admitted that they did not adequately protect the privacy of consumers and "tightened up" their systems to address the issue.[27] While we are thankful that Google acknowledged that there was an issue and took responsible measures to address it, we would like to know how Google plans to prevent Google Glass from unintentionally collecting data about the user/non-user without consent?

By the time the subcommittee had responded, it was in the context of a series of criticisms and clarifications that legal scholars like Schwartz and Solove were already making about the importance of understanding how privacy should be understood and framed within societies increasingly dominated by surveillance technologies, which included everything from Google Glass to drones, and educational privacy in student data collection (Shwartz & Solove, 2012; Solove, 2011). Their memorandum regarding privacy and American Law Institute projects[28] specifically addressed the intentional emotional harm and impact of those affected by invasions of privacy. By the writing of this chapter, U.S. law has not caught up with the rapid proliferation of surveillance technologies, as the lines between emotional harm as a result of invasions of privacy through social media and Internet reproduction and distribution of photos, videos, and information and the loss of rights to privacy in public continue to blur.

ANALYZING CLASS THROUGH GLASS

Unlike Google's characteristic approach to launching products for "free" or inexpensively, Sergey Brin, one of Google's founders, engaged a formula

---

[24] A. Schatz and A. Efrati (2010, November 2). FCC investigating Google data collection. *Wall Street Journal.* http://www.wsj.com/articles/SB10001424052748704804504575606831614327598.

[25] See footnote 24.

[26] B. Sasso. "Google pays $7 million to settle Wi-Fi snooping charges," *The Hill* (2013, March 12), http://perma.cc/KS6U-KE8W.

[27] See footnote 26.

[28] See: https://law.duke.edu/sites/default/files/images/centers/judicialstudies/Reworking_Info_Privacy_Law.pdf.

directly out of the tastemaker product launch playbook for high-end goods: he seeded Glass with the White American social and cultural elite. Google's marketing team developed a Google Glass Explorers program that allowed the public to test the product and provide feedback to the design team for $1500. At that price point, only those with the disposable income or adequate resources were able to become Glass Explorers. As a result of its price for participation, Google Glass was marketed as a high-end luxury item, its arrival showcased on the runway at a *New York Fashion Week* launch by designer Diane von Furstenberg (Figure 9.5).

Figure 9.5 From left, Sergey Brin, cofounder of Google; Diane von Furstenberg, an early adopter of Google Glass, and Yvan Mispelaere. Credit: Frazer Harrison/Getty Images for Mercedes-Benz. Source: *New York Times*.

The blogosphere responded to Google Glass with *White Men Wearing Google Glass*, a Tumblr dedicated to joking about another technology development serving the interests of white men (Figure 9.6).

The commentary, as evinced in its photo posting of Prince Charles of Great Britain wearing Glass along with the caption, "In its favor, if Google Glass didn't exist, all these Silicon Valley guys would be having affairs or buying unsuitable motorbikes," could be a dig at Google executives who both have had very public affairs and divorces (Bilton, 2015) and are experimenting

Figure 9.6 Prince Charles of Great Britain wearing Google Glass. Source: http://whitemenwearinggoogleglass.tumblr.com/.

with things like autonomous motorcycles.[29] But the context of "Google Glass on White Men" is situated by the ways in which surveillance technologies are rooted in celebration the freedoms afforded White Americans, and the ultimate role that surveillance technologies have provided in the protection of property for the wealthy (Blakely & Snyder, 1997; Flusty, 1994) who have historically been, and contemporarily are, White men. The normalization of surveillance of "the other" by White Americans, in particular, is a product of the ways in which White property, White interests, and even Whiteness itself, can and should be protected at all costs (Harris, 1995). Surveillance technologies like Google Glass, on the faces of White men in the case of the Tumblr, are illustrative of the ways in which Whiteness and panoptic control over "the other" are not only normalized, but celebrated uncritically in popular culture and in the launch of the product to the public.

SURVEYING EMOTIONS AND SPACE THROUGH DESIGN

While tremendous attention has been focused on facial recognition and surveillance, a small company called Emotient.com has developed a new

---

[29] E. Ackerman, "Google wants option to test autonomous motorcycles and trucks in California," http://spectrum.ieee.org/cars-that-think/transportation/self-driving/google-autonomous-motorcycles-and-trucks-in-california.

Google Glass application that captures users' emotions and relays them back to advertisers in real-time.[30] Emotient's software scans changes in facial expressions and mood, with the goal of helping retailers understand how consumers are responding to new campaigns and offers. The company alleges that through emotion tracking it can help medical practitioners intervene sooner and save money and time in protracted medical recovery from surgeries due to mental health crises like depression. But investment in emotion-recognition software is about refining advertising impact, and given Google's current profit model and status as the world's largest advertising agency, it is logical to read Google's investment in Emotient. com as directly related to their ability to better serve up ads to consumers, particularly through software that can better improve its click-through rates (Nissenbaum and Introna, 2004; Noble, 2013b).[31] But what are the ethics of commodifying consumer emotions? Currently, Emotient says it aggregates users' data and anonymizes it before communicating back to its advertiser clients and so there are no privacy issues for individual users.[32]

The politics and ethics of commodification of biodata, which we argue includes emotion-recognition, has become an important consideration to digital media and information scholars. Bronwyn Parry's (2004) research on the convergence of technical, social, and regulatory interests governing trade and speculation on biodata is an extensive analysis of the ways that surveillance is manifest at the level of genes and other biomarkers. Concerns over the politics of biodata have been raised in new lawsuits over the ownership of embryonic cells and reproductive matter, including the emergence of markets for genetically designed babies (Roberts, 2011). In the United States, the National Science and Technology Council established a subcommittee to investigate protections for the public in the collection of biodata, which is largely considered to be very private data, or "Personally Identifiable Information" that is legally protected and carries criminal penalties for infringements (Sarpu, 2014). Biodata is a complicated new frontier of data collection and surveillance, however, on the Internet. In this realm, legal protections with regard to the use

---

[30] See: http://www.fastcompany.com/3027342/fast-feed/this-google-glass-app-will-detect-your-emotions-then-relay-them-back-to-retailers.

[31] Google's advertising program, *AdWords*, generates profits for the company by encouraging consumers to click on ads, or websites that appear to be informational, even when they are advertisements. It also serves up ads and links that are part of their vertical and horizontal business holdings, which is known as Google Bias.

[32] See footnote 31.

of bioidentity markers are murky. As new products are flooding the marketplace—from the aforementioned Fitbit[33] exercise monitors and Apple Watch, to medical devices. Google Glass was, and is, uniquely positioned to benefit from the commodification of biodata through its vertical and horizontal business partnerships that are perfecting emotional and biodata surveillance as a site for the expansion of capital. Users are beginning to recognize these intrusions, particularly in the use of facial recognition software used by companies like Google and Facebook, and are unhappy and responding negatively to what they perceive as invasions of privacy (Aguado, 2012; Sarpu, 2014).

Understanding emotionality, however, in the context of the design of surveillance technologies requires a nuanced reading of space, both real and imagined. Space and spatiality of life has been theorized as an important site of social life and conflict (Harvey, 1973; Koskela, 2000; Lefebvre, 1991; Massey, 1994; Rose, 1993; Soja, 1996). Koskela (2000) has extensively theorized the ways in which surveillance is changing the nature of space, understandings of private versus public space, and its contestations, which she characterizes in three distinct and overlapping ways: space as a container, power-space, and emotional space. Although her work is primarily focused on surveillance cameras in public spaces, rather than on individual wearers of technologies like Google Glass, her insights about the design of these artifacts are important and relevant:

What is characteristic of surveillance design is its paradoxicality: forms are at the same time transparent and opaque. While everything (and everyone) under surveillance is becoming more visible, the forces (and potential helpers) behind this surveillance are becoming less so.

(Koskela, 2000, p. 250)

On one hand, the design of wearable technologies like Glass is made completely visible, on the face of the user; but paradoxically, it is the unknown ways in which the Google gaze is enveloping everything and everyone in its line of sight, and in service of what, and for whom, at what cost? Google's fundamental business model is based on commodifying the gaze, or "pay-per-gaze" advertising (Sorg, 2013). Space, then, and the lack of an ability to defend it from encroachment, is a social construct that is laden with power relations–a concept similarly embraced by information and technology scholars who argue that technological spaces, like cyberspace, as socially constructed and exist in service of White male power (Brock, 2011, 2009; Daniels, 2012; Kendall, 2002; Kettrey & Laster, 2014; Noble, 2013a).

---

[33] https://www.fitbit.com.

Who is contained in the surveyed space is important, and technology is rapidly shifting the landscape, prompting greater need to understand how power is exercised through the defining and erosions of space through racialized and gendered practices. For example, those who often control surveillance policy and technology design are men, and, through gendered labor, the interests and concerns of women are often excluded. This is apparent in practices like revenge porn,[34] and the ways that explicit footage of women has been taken without consent in restrooms, changing rooms, topless beaches where women are significantly more likely to be sexually surveyed as objects of a male photographic or video gaze (Koskela, 2000). Wajcman (1991) extensively discusses the "masculine culture" of technological practices and designs, and details the ways that technologies work to the benefit of men, and how gendered power relations are often reproduced through technological practices.

Surveillance is fundamentally intertwined with emotion. Koskela (2000) emphasizes the ways in which power and emotions are inevitably enmeshed, and how emotional space "is a space 'below the threshold at which visibility begins'" (de Certeau, 1984, p. 93, quoted in Gregory, 1994, p. 301). She writes, "the variety of feelings surveillance evokes is enormous: those being watched may feel guilty for no reason, embarrassed or uneasy, irritated or angry, or fearful; they may also feel secure and safe" (Koskela, 2000, p. 257). Because emotionality is often a feminized concept (Rose, 1993), it is generally disregarded in the realm of technology's masculine culture, thus, as Koskela argues, emotion is an undervalued experience. In many ways, it is those who are more likely to be surveyed who are responding with intense emotion to projects like Google Glass, because the social experience of being seen and watched, or recognized and searched through a search engine in real-time, renders people more vulnerable than the watcher.

CONCLUSION: THE FUTURE OF GLASS

The brief life of Google Glass eventually ended with the retirement of the project after resounding public backlash, although Google Glass 2.0 is currently under development and scheduled for a potential rerelease in the next year.[35] Eyewear giant Luxottica recently announced to its shareholders that

---

[34] See: Dr Mary Anne Franks, "Criminalizing revenge porn: a quick guide" at http://www.endrevengeporn.org/guide-to-legislation/.

[35] See: "Google Glass lives! version two coming soon" by Eric Mack. http://www.forbes.com/sites/ericmack/2015/04/24/google-glass-version-2-0-coming-soon-says-eyewear-giant-luxottica/.

the project is reemerging as a more fashionable product that consumers will enjoy.[36] The Wall Street Journal reported, "Glass chief Ivy Ross in January said the updated gadget will be cheaper and have longer battery life, improved sound quality, and a better display. Google is also trying to tackle the social stigma of Glass by pairing the device with more familiar types of eyewear."[37] In many ways, Glass is an important technological artifact that underscores the important interplay between emotions, technology, and design. The company is taking on these challenges in order to redeploy.

Unlike many of its previous sociotechnical triumphs, Google Glass was not a resounding success for the company. Rather than propel the product, its exclusivity actually seemed to contribute to hostilities that non-Glass users visited upon those who wore the technology in public. The non-Glass public rejected Glass both on the grounds of its technological (read: surveillance) capabilities, and their lack of power to control how Glass was being used upon them, as well as on larger social grounds, based on the class divisions that Google Glass came to represent in already highly stratified areas such as San Francisco and Seattle, two cities where the economic gulf between technology haves and have-nots only seems to continue to grow.

In the field of robotics (among others), the notion of the "Uncanny Valley" describes the point at which a human's positive emotional response to a robotic entity (Mori, 2012) dissipates once the humanness of that entity surpasses a certain point of likeness.[38] In other words, human positive affect dips precipitously (the "valley" in question) once the robotic entity becomes too human. The Uncanny Valley concept, therefore, has huge implications for technological design. Perhaps the Google Glass experiment has discovered a new sort of Uncanny Valley, this time, in relation to wearable technology and our own comfort with self- and external technosurveillance. At a time when countless scholars and critics have questioned at what point people would resist the mass trade-off of personal information autonomy for the technological convenience and romance of sophisticated gadgetry, Google Glass seemed to have found itself in that very precipitous dip, the site at which the public rejected its convenience and techno thrills—vociferously and, sometimes, even violently.

---

[36] See footnote 35.
[37] See: "Italian eyewear maker Luxottica working on new version of Google Glass, CEO Says" by Manuela Mesco at: http://blogs.wsj.com/digits/2015/04/24/italian-eyewear-maker-luxottica-working-on-new-version-of-google-glass-ceo-says/.
[38] See: English translated article and interview with Mori at: http://spectrum.ieee.org/automaton/robotics/humanoids/the-uncanny-valley.

Importantly, in the case of Glass, the gaze of the machine was turned toward those who were not Glass-enabled. Does the Uncanny Valley for wearables find its tipping point where the surveillance of the device turns outward? How would Glass have been received had its camera and general, highly distinctive design been less obvious to those within its gaze (see Luxottica's project as described above)? Or was the limited rollout of Glass into markets that were already engaged in deep economic and class-based rifts (due in no small part to the effect of technohubs on the cities nearby), part of its demise? Very few trade and popular press articles have focused on the failure of Glass along these dimensions, and yet the surveillance and class-based aspects of Google Glass are fundamental to an accurate rendering of the product's trajectory and the public's emotional response to this product. In this chapter, we offer this alternate view: one which foregrounds dimensions of surveillance and economics, class and resistance, in the face of unending rollouts of new wearable products designed to integrate seamlessly with everyday life–for those, of course, who can afford them. Ultimately, we believe more nuanced, intersectional analyses of power along race, class, and gender must be at the forefront of future research on wearable technologies. Our goal is to raise important critiques of the commodification of emotions, and the expansion of the surveillance state vis-à-vis Google's increasing and unrivaled information empire, the longstanding social costs of which have yet to be fully articulated.

REFERENCES

Aguado, C. (2012). *Comment, Facebook or face bank?* 32 LOY. L.A. ENT. L. REV. 187, 195.
Ahmed, N. (2015, January 22). *Why Google made the NSA.* Retrieved September 28, 2015, from https://medium.com/insurge-intelligence/why-google-made-the-nsa-2a80584c9c1
Andrejevic, M. (2007). Surveillance in the digital enclosure. *The Communication Review*, *10*(4), 295–317.
Bilton, N. (2015, February). Why Google Glass broke. *The New York Times.* http://www.nytimes.com/2015/02/05/style/why-google-glass-broke.html. Accessed 10.6.2015.
Blakely, E. J., & Snyder, M. G. (1997). Divided we fall: Gated and walled communities in the 262 video-surveillance and the changing nature of urban space United States. In N. Ellin (Ed.), *The architecture of fear* (pp. 85–99). New York: Princeton Architectural Press.
Brock, A. (2009). Life on the wire: Deconstructing race on the Internet. *Information, Communication & Society*, *12*(3), 344–363.
Brock, A. (2011). Beyond the pale: The Blackbird web browser's critical reception. *New Media and Society*, *13*(7), 1085–1103.
Crampton, J., & Krygier, J. (2006). An introduction to critical cartography. *ACME: An International E-Journal for Critical Geographies*, *4*(1), 11–33.
Crawford, M. (1992). The world in a shopping mall. In M. Sorkin (Ed.), *Variations on a theme park: The new American city and the end of public space* (pp. 3–30). New York: Noonday Press.

Daniels, J. (2012). Race and racism in Internet studies: A review and critique. *New Media and Society*, *15*(5), 695–719. http://dx.doi.org/10.1177/1461444812462849.
de Certeau, M. (1984). *The practice of everyday life*. Berkeley, CA: University of California Press.
Flusty, S. (1994). *Building paranoia: The proliferation of interdictory space and the erosion of spatial justice*. Los Angeles, CA: Los Angeles Forum for Architecture and Urban Design.
Foucault, M. (1977). *Discipline and punish: The birth of a prison*. (A. Sheridan, Trans.). London: Penguin Books.
Goldman, A. (2013). *The "Google shuttle effect:" Gentrification and San Francisco's dot com boom 2.0*. Unpublished master's thesis. University of California, Berkeley.
Goss, J. (1995). Marketing the new marketing: The strategic discourse of geodemographic information systems. In J. Pickles (Ed.), *Ground truth: The social implications of geographic information systems* (pp. 130–170). New York: Guilford.
Greenberg, A. (2014). *Wi-Fi detector lets you create a Glasshole-free zone*. http://www.wired.co.uk/news/archive/2014-06/04/ban-glassholes-wifi. Accessed 8.6.2015.
Gregory, D. (1994). *Geographical imaginations*. Oxford: Blackwell.
Harris, C. (1995). Whiteness as property. In K. Crenshaw, et al. (Eds.), *Critical race theory: The key writings that informed the movement*. New York: The New Press.
Harvey, D. (1973). *Social justice and the city*. London: Arnold.
Harvey, F., Kwan, M., & Pavlovskaya, M. (2005). Introduction: Critical GIS. *Cartographica*, *40*(4), 1–4.
Judd, D. R. (1995). The rise of the new walled cities. In H. Liggett & D. C. Perry (Eds.), *Spatial practices. Critical explorations in social/spatial theory* (pp. 144–166). Thousand Oaks, CA: Sage.
Kendall, L. (2002). *Hanging out in the virtual pub: Masculinities and relationships online*. Berkeley, CA: University of California Press.
Kettrey, H. H., & Laster, W. N. (2014). Staking territory in the "world white web": An exploration of the roles of overt and color-blind racism in maintaining racial boundaries on a popular web site. *Social Currents*, *1*(3), 257–274.
Kloc, J. (2014). Tech boom forces a ruthless gentrification in San Francisco. *Newsweek*. http://www.newsweek.com/2014/04/25/tech-boom-forces-ruthless-gentrification-san-francisco-248135.html. Accessed 11.6.2015.
Koskela, H. (2000). The gaze without eyes': Video-surveillance and the changing nature of urban space. *Progress in Human Geography*, *24*(2), 243–265.
Lefebvre, H. (1991). *The production of space* (D. Nicholson-Smith, Trans.). Oxford: Blackwell.
Mann, S., Nolan, J., & Wellman, B. (2003). Sousveillance: Inventing and using wearable computing devices for data collection in surveillance environments. *Surveillance and Society*, *1*(3), 331–355.
Massey, D. (1994). *Space, place and gender*. Cambridge: Polity Press.
Matyszczyk, C. (2014, April 8). *72 percent say no to Google Glass because of privacy*. http://www.cnet.com/news/72-percent-say-no-to-google-glass-because-of-privacy/. Accessed 8.6.2015.
Miller, C. C. (2013). Google searches for style. *The New York Times*. http://www.nytimes.com/2013/02/21/technology/google-looks-to-make-its-computer-glasses-stylish.html. Accessed 8.8.2015.
Mori, M. (2012). *The Uncanny valley*. http://spectrum.ieee.org/automaton/robotics/humanoids/the-uncanny-valley. Accessed 10.6.2015.
Nissenbaum, H., & Introna, L. (2004). Shaping the web: Why the politics of search engines matters. In V. V. Gehring (Ed.), *The Internet in public life* (pp. 7–27). Lanham, MD: Rowman & Littlefield.
Noble, S. U. (2011). Geographic information systems: A critical look at the commercialization of public information. *Human Geography: A New Radical Journal*, *4*(3), 88–105.

Noble, S. U. (2013a). Google search: Hyper-visibility as a means of rendering black women and girls invisible. *InVisible Culture*, 19.
Noble, S. U. (December, 2013b). Search engine bias/Google bias. In K. Harvey (Ed.), *Encyclopedia of social media and politics*. Thousand Oaks, CA: SAGE Reference.
Pacey, A. (1983). *The culture of technology*. Cambridge, MA: MIT Press.
Parry, B. (2004). Introduction. In *Trading the genome: Investigating the commodification of bioinformation* (pp. 1–11). New York: Columbia University Press.
Roberts, D. E. (2011). *Fatal invention: How science, politics, and big business re-create race in the twenty-first century*. New York: New Press.
Roberts, S. (2015, 2016). Commercial content moderation: Digital laborers' dirty work. In S. Umoja Noble & B. Tynes (Eds.), *The intersectional Internet. Digital formations series* (S. Jones, series ed.). New York: Peter Lang.
Rose, G. (1993). *Feminism and geography. The limits of geographical knowledge*. Minneapolis, MN: University of Minnesota Press.
Sarpu, B. (2014). Google: The endemic threat to privacy. *Journal of High Technology Law*, *15*(1), 63–101.
Schiller, H. (1996). *Information inequality: The deepening social crisis in America*. New York, NY: Routledge.
Schuurman, N. (2000). Trouble in the heartland: GIS and its critics in the 1990s. *Progress in Human Geography*, *24*, 569–590.
Shields, R. (1989). Social spatialization and the built environment: The West Edmonton Mall. *Environment and Planning D: Society and Space*, 7, 147–164.
Shwartz, P. M. & Solove, D. J. (2012). *Reworking information privacy law: A memorandum regarding future ALI projects about information privacy law*. https://law.duke.edu/sites/default/files/images/centers/judicialstudies/Reworking_Info_Privacy_Law.pdf. Accessed 10.6.2015.
Soja, E. W. (1996). *Thirdspace. Journeys to Los Angeles and other real-and-imagined places*. Cambridge, MA: Blackwell.
Solove, D. J. (2007). *The future of reputation: Gossip, rumor, and privacy on the Internet*. New Haven: Yale University Press.
Solove, D. J. (2011). Why privacy matters when you have 'nothing' to hide. *The Chronicle of Higher Education*. http://chronicle.com/article/Why-Privacy-Matters-Even-if/127461/. Accessed 8.8.2014.
Sorg, L. (2013, October 9). A crack in Google Glass: Wearable technology's glassault on privacy. *Indy Week*. http://perma.cc/KA9N-2KUJ.
Sullivan, G. (2014, May 29). Google statistics show Silicon Valley has a diversity problem. *The Washington Post*. http://www.washingtonpost.com/news/morning-mix/wp/2014/05/29/most-google-employees-are-white-men-where-are-allthewomen/.
Vaidhyanathan, S. (2011). *The Googlization of everything: (And why we should worry)*. Berkeley: University of California Press.
Wajcman, J. (1991). *Feminism confronts technology*. University Park, PA: Pennsylvania State University Press.
Wilson, J. Q., & Kelling, G. L. (1982). Broken windows. *Atlantic Monthly*, *249*(3), 29–38.
Winner, L. (1986). Do artifacts have politics? In *The whale and the reactor: A search for limits in an age of high technology* (pp. 19–39). Chicago, IL: University of Chicago Press.

CHAPTER 10

Designing Emotions: *Deliver the Nets*, Eradicate Malaria in Africa, and Feel Good?

Ergin Bulut[a], Robert Mejia[b]
[a]Koc University, Istanbul, Turkey
[b]SUNY Brockport, Brockport, NY, USA

DESIGNING THE EMOTIONS OF COMMUNICATIVE CAPITALISM

Although digital games capture our society's imagination and produce a public discussion mostly in relation to events of mass violence, the ways in which play has been put to work for serious purposes (such as education, military training, budgeting, job applications) have barely registered in critical analyses. This is unfortunate for we have argued elsewhere that video games can operate within the context of "communicative capitalism" (Bulut, Mejia, & McCarthy, 2014). Jodi Dean (2009) defines communicative capitalism as "the materialization of ideals of inclusion and participation in information, entertainment, and communication technologies in ways that capture resistance and intensify global capitalism" (p. 2). As it pertains to video games, we argue that the game design can erase our social memory and obscure the political economic environment of which gaming is a part.

For this analysis, our interest is in the game *Deliver the Nets*, which can be situated within the rise of the serious games movement. This is a game design movement that is motivated by a faith in the power of digital technologies to resolve social problems and fix what McGonigal (2011) has described as a "broken reality." This discourse holds that digital games have immense potential to resolve social problems including those of government and education (Gee, 2005). Diane Tucker (2012), director of Woodrow Wilson Center's Serious Games Initiative, is worth quoting at length to illustrate the optimism of the serious games movement:

> As today's policy of challenges become more complex, it has become clear that American media—online news, television, radio, newspapers, and magazines—are not up to the task of explaining the problems underlying them or providing

citizens with all the information they need to engage in public conversations. But one new medium—videogames—may well fill the gap. By their very nature, videogames can engage players in ways that enable players to make their way through the intricacies of policy problems. As players begin to understand them in all their complexity, games may well help their governments forge solutions.

While this faith in what is often called techno-utopianism has existed since the emergence of modernity—beginning with a faith in the mechanical, then the electrical, and so on (Carey & Quirk, 2009)—the ways in which video games recruit the player to become a part of the narrative trajectory warrants particular attention (Galloway, 2006). This is because video games have unique technological affordances for creating intense simulated emotions, engaging players with virtual worlds and thus ultimately "reassert, rehearse, and reinforce" certain types of subjectivities including that of "worker-consumer and soldier-citizen" (Dyer-Witheford & de Peuter, 2009, p. xiv).

Indeed, as the ideal commodities of informational capitalism, digital games reinforce the Western subject's emotions of autonomy and altruism since they "embody the liberal ideals of individual choice and agency. One can play what one wants to play at any time and in any place if one can afford it" (Kerr, 2006, p. 1). It is precisely these emotional affordances of autonomy and freedom that create the market for serious games through which our society increasingly lends itself toward gamification of real life situations. Hence, drawing on a theoretical background that connects visuality to the construction of power relations, this chapter provides a representational analysis of *Deliver the Nets* and ultimately argues that, as a serious game, *Deliver the Nets* portrays Africa as a faceless continent outside of history. It provides an easy route to the eradication of serious social problems, thereby creating emotional relief on the side of the player, who is no longer obliged to tackle the question of underdevelopment as a political matter but rather perceives it as a playful one.

SERIOUS GAMES IS SERIOUS WORK

Before analyzing the design aspects of *Deliver the Nets*, it is important to historicize it as a "serious game." While the idea of a serious game might sound like an oxymoron, digital games are increasingly transforming play from being a "wasteful" activity into "as a site of social, cultural, and political productivity" (Bulut et al., 2014, p. 4). Despite the seeming novelty of serious games, one needs to note that Clark Abt (1987) had already imagined the "serious" potentials of games in his book *Serious Games*. Indeed, video games today are designed for a variety of purposes, such as to help "teach slavery and American Civil War, help disabled children navigate real-world environments, educate the public about NASA's missions, train military personnel, and learn to balance the federal budget at a time of financial crisis" (Bulut et al., 2014, p. 5).

The emotional and affective reach of digital games is perhaps best captured by Alexander Galloway, who describes games as "an active medium that requires constant physical input by the player: action, doing, pressing buttons, controlling" (2006, p. 83). Indeed, "the gamer is not simply playing this or that historical simulation. The gamer is instead learning, internalizing, and becoming intimate with a massive, multipart, global algorithm" (Galloway, 2006, p. 91). As Galloway argues, "video games are, at their structural core, in direct synchronization with the political realities of the information age" and they "teach structures of thought" (Galloway, 2006, p. 91).

How, then, does one understand serious games *vis-à-vis* Galloway's description? What kind of emotional work do serious games perform? First of all, serious games involve a cultural experience that goes beyond pure entertainment. They are still games but have a more hybrid form, involving both play and seriousness. These games "allow learners to experience situations that are impossible in the real world for reasons of safety, cost, time, etc., but they are also claiming to have positive impacts on the players' development of a number of different skills" (Susi, Johannesson, & Backlund, 2007, p. 1). One can, for instance, learn to sympathize with 2.5 million refugees in Darfur by playing "Darfur is Dying."

In this sense, the ways in which serious games are designed to create emotions that one cannot "feel" in "real life" is important because citizens are increasingly forced to draw on digital technologies for education. Yet, while research regarding the gamification has focused much on education (Gee, 2007; Kirriemuir & McFarlane, 2004), production of serious games is by no means limited to the realm of education. It is very well known, for instance, that "the military is the major user of serious games" and this has been the case since *Atari* developed *Army Battlezone* [in the 1980s] for military training (Susi et al., 2007, p. 11). Perhaps, the most famous digital venue for military training and serious gaming has been *America's Army*. "Downloaded more than 42.6 million times, more than any other war game" (PBS Frontline, 2010), the game is "free to play online, courtesy of a publicly funded, multi-million-dollar investment by the

US Department of Defense" (Dyer-Witheford & de Peuter, 2009, p. xiii). It has proven to be cost-efficient and practical, too, in that "it has helped the Army to recruit soldiers at 15% of the cost of other recruiting volunteers." Indeed, "30% of Americans between the ages 16–24 claim to have learned some of what they know about the Army from this game" (Susi et al., 2007, p. 11). Other than military, there are government games which involve "different types of crisis management, for instance, dealing with terrorist attacks, disease outbreaks, biohazards, health care policy issues, city planning, traffic control, firefighting, budget balancing, ethics training, and defensive driving" (Susi et al., 2007, p. 11).

Contrasted with *America's Army*, the design of digital games that cultivate emotions of sympathy are to be praised. And yet, our earlier work on philanthropic video games suggests that the causal relationship between play and philanthropy ought to be approached cautiously (Bulut et al., 2014). Such a causal relationship implies an instrumental view of technology, which does not take history or social totality into consideration. Moreover, an instrumental view of technology treats video games as things with identifiable cognitive affects rather than social practices which involve more of an interaction and meaning-making (Arnseth, 2006; Bulut et al., 2014). If games are worlds of design that involve particular sociocultural practices and political economic substructures, then they must also serve to produce certain emotions or form particular subjectivities that cannot be thought independently of that totality in which they are produced, played, and made sense of.

In this respect, as far as video games research is concerned, we need to transcend the dualism of condemnatory/celebratory approaches. "Unlike earlier generations of media-effects perspectives that emphasized individual psychologies," there needs to be more emphasis on "social structures, corporate contexts, and institutional forces" (Dyer-Witheford & de Peuter, 2009, p. xxvii). In other words, history matters. The historical moment into which any particular game, *Deliver the Nets* in our case, is born needs to be well understood. This historical moment—"the political reality of information age" in the words of Galloway—within which *Deliver the Nets* emerges, we believe, is that of what Roy (2010) calls "millennial development."

This millennial development is closely tied to discourse of global citizenship with an unquestionable investment in the power of technology, which becomes a significant asset in the construction of neoliberal hegemony that rests not only on economic strategies but the work of the cultural, the mundane, and now what counts as fun. It is a milieu of development, actively involving simulation and gamification of ideals of progress, ignoring geographical and historical context for development. As Neubauer (2011) has argued, it constitutes a kind of "information populism" that derives its power from the discourse and spirit of networks with unlimited power to overcome social problems. For Fisher (2010), our contemporary historical moment is one in which technology is rearticulated in the new hegemonic project by legitimizing the withdrawal of the state from social policies as well as creating a new regime of truth where there is no room or patience for learning the history of underdevelopment. Within this context, underdevelopment is constructed as a historical accident through a visual matrix where certain geographies are put outside history and it is what we turn to analyze next by looking at *Deliver the Nets*.

DELIVER THE NETS: ERADICATE MALARIA, REIFY AFRICA

As an educational game, *Deliver the Nets* aims to eradicate malaria and its popularity is to be commended. Even though the United Nations Foundation had been struggling to fight malaria for some time, it was only when Rick Reilly from *Sports Illustrated* wrote about malaria that there emerged a popular interest, which ultimately created the "Nothing but Nets" campaign. In terms of its discourse, the campaign is designed very democratically, because "Nothing But Nets" "allows everyone from students to CEOs, bishops to basketball players, to join the fight against malaria by giving $10 to send a net and save a life" (http://www.nothingbutnets.net/nbn/worldmalariaday.html). Fitting to the shared governance spirit of the "millennial development" (Roy, 2010), the campaign involves various partners including "Bill & Melinda Gates Foundation, NBA Cares, The People of The United Methodist Church, Major League Soccer's MLS WORKS, the Union for Reform Judaism, and Junior Chamber International." Echoing the emotional atmosphere of our age where the digital renders anything possible, contributors and players are told that "Nothing But Nets is powered by your passion; it works because you do." Working with partners including UNICEF, UN Refugee Agency, and World Health Organization, the campaign is designed to raise funds to "purchase and distribute bed nets during national vaccination campaigns." The rationale for the project is as follows:

> Refugees in Africa have fled their native countries because of political and economic strife, ethnic tensions, and natural disasters. They traveled into the unknown with little more than the clothes on their backs in hopes of a better, safer life. Some refugees settle in camps for months, some for years. But their stay can be cut short

with just one deadly mosquito bite. Refugees are particularly vulnerable to malaria—in fact, malaria is the number one killer of refugees in Africa[1].
(http://www.nothingbutnets.net/nets-save-lives/refugees.html)

The image of the refugee and the whole continent of Africa constructed here is one of complete vulnerability. Yet, the refugees are not helpless "thanks to" first world intervention. Focusing especially on Sub-Saharan Africa, the campaign intends to educate the global society about malaria, the regions that the campaign targets, the impact of the campaign and more importantly ways to engage and fight malaria: direct donations beginning from $10 USD; engaging in sports; "starting creative challenges to raise money" such as "designing bracelets, hosting a bake sale, or putting a concert"; reaching out to members of congress; "engage with Team Bzzzkill on Twitter and Facebook"; or play *Deliver the Nets* and "experience net delivery first hand[2]" (http://nothingbutnets.net/createfornets/).

In order to play *Deliver the Nets*, the player-citizen undertakes the role of a male driver (not White, but Black) who rides a motorcycle to deliver nets. The player is racing against time, which is important in terms of how many nets one can possibly deliver. As one plays the game, one comes across different characters such as a smiling Black girl or a Black woman holding a kid. As we played this game, the only male character we encountered was the

---

[1] At the time of the writing, this link worked but as of today, it cannot be located. For a glimpse of the game, readers can see http://www.nothingbutnets.net/its-easy-to-help/game.html.
[2] http://www.nothingbutnets.net/nets-save-lives/.

person who rode the motorcycle. The emotions we end up having with respect to "helpless and faceless refugees" are further intensified by the spatial configuration of the game scene. The player rides the motorcycle through residences that are not "modern." There are small ponds, rocks, or trees around which the player needs to steer the motorcycle and thus be active. There are also some motionless animals including zebras and goats. There is no sound in the game. As the sun goes down, there emerge mosquitoes—the very reason for malaria.

What is important to highlight as far as game design and cultivation of emotions is that the people to whom we deliver these nets seem to exist in a world of stasis. Lacking any political agency, they are motionless, while it is us who are able to move and *Deliver the Nets*. If the player is successful enough to run out of nets, a UN truck emerges where you press "X" button to load more nets to deliver, thus opposing the world of bounty and cornucopia to the world of lack and immobility. At the end of the game, you reach one of those "primitive" residences where there is an African woman sitting inside. The quarry of the gamer's quest has been located—poverty submits itself to enlightened action. Through virtual play, we "feel" what it means to deliver nets as well as what it means to live under the threat of malaria.

This representation of Africa as a static premodern continent matters for, especially since the advent of smart phones, visual culture's impact on subjectivity, the cultivation of emotions, and formation of power relations has reached an extraordinary phase especially in the advanced capitalist West, which enjoys the wealth created through the exploitation of Africa during colonialism and still relies on uneven relations of development. Living media-saturated lives is especially the case in advanced capitalist geographies, where the pace of innovation has facilitated the emergence of new information and communication technologies with an intensified explosion of images flowing across a plethora of screens. This is not to say that visuality was never important before (Ewen, 1988; Rose, 2001). What we are arguing instead, here, is that it is rather the explosion and speed within which these images get circulated, put into hybridized combination and made sense of.

In order to make sense of these images and the way they are produced and circulated, a critical approach to any text or image "has to address questions of cultural meaning and power" (Rose, 2001, p. 3). The phrase "cultural meaning" is important to underline since culture is the embodiment of material practices through which we make meaning, construct a reality, and produce an understanding of the world and ourselves. In this respect, visuality and interactivity are but one part of this construction. As John Berger insightfully argues, seeing is *relational and active* in that "we never look at just one thing; we are always looking at the relation between things and ourselves. Our vision is continually active, continually moving, continually holding things in a circle around itself, constituting what is present to us we are" (Berger, 1972, p. 9). That is, to look is to actively engage with a particular object, and thus to construct meaning and reality; we attempt to make a sense of the world and understand it as we look at images and this "entails, among other things, thinking about how they offer very particular visions of social categories such as class, gender, race, sexuality, ablebodiedness, and so on" (Rose, 2001, p. 11).

Yet, as Gillian Rose further argues, it is not just the images that matter but how "it is seen by particular spectators who look in particular ways" (Rose, 2001, p. 12). In other words, as Rose claims with reference to Berger's analysis of the nude women in Western art, the images not only represent what they show but also construct a reality and invite certain ways of seeing. In that sense, the relationship between the image and the subject is such that it positions the viewer and constructs positionality. Furthermore, the images and representations are there as part of a historical totality, with its practices of production, consumption, and constitution of identity. Here, we use the notion of totality not as a closed system but an open one, as used by Fredric Jameson. That is, totality is "composed of differences and discontinuities, residual and emergent forces, antinomies and contradictions, and anomalies"; it is "meant to function as a prescription to strive constantly to relate and connect, to situate and interpret each object or phenomenon in the context of those social and historical forces that shape and enable it" (Hardt & Weeks, 2000, pp. 21–22).

Therefore, we argue, we need to take into account that *Deliver the Nets* is emerging in a historical moment within which practices of ludic governance are gaining momentum, either in the form of concerts or digital games. Once we understand the context, do we need to probe the representational and emotional work produced by *Deliver the Nets*? While we are not proposing to measure the "effects" of the game, it is possible to formulate some thoughts on what *Deliver the Nets* and the images it presents mean.

To begin with, in *Deliver the Nets*, Africa emerges as a reified geography. It is a place where we see Black people, motionless, waiting for help. There is even one little girl who has opened her hand, posited to be asking for help. This dichotomy of mobile male vs. motionless women and kids tells us a couple of things. The attribution of being mobile to a male character demonstrates the gendered character of the game. Additionally, who is giving what to whom is a whole matter of power relations that not only goes back to the roots of colonialism but also is rooted within contemporary societies divided along class, race, and gender lines. This is where acts of philanthropy come into play nationally and in forms of global governance. While philanthropy undoubtedly helps the needy, it is not the only function it fulfills. As it has been strongly maintained, "a society relying too exclusively on philanthropy can easily fall into moral traps because it creates the fiction of a self-sufficient giver and an insufficient receiver" (Eikenberry & Nickel, n.d., p. 12). In a similar vein, a discursive construction of marginality serves to portray the social inequalities as natural and unavoidable.

If we go back to the representation in *Deliver the Nets*, it would be appropriate to argue that the reification of Africa as a continent that is outside history is troubling. The question of why Africa is, almost as a whole continent, in a position to be suffering from malaria in the very first place is never asked. In that sense, the game, images, and discourse of the website constructs Africa as a location outside history, disregarding its colonial past and anticolonial struggles. As the anticolonialist cultural critic, Césaire (1955/2000) wrote:

> They talk to me about progress, about "achievements," diseases cured, improved standards of living.
> [...]
> I am talking about natural economies that have been disrupted—harmonious and viable economies adapted to the indigenous population—about food crops destroyed, malnutrition permanently introduced, agricultural development oriented solely toward the benefit of the metropolitan countries; about the looting of products, the looting of raw materials (pp. 42-43).

In other words, *Deliver the Nets* perpetuates the same (neo)colonialist rhetoric that Césaire diagnosed 60 years ago; this is that of a manufactured social amnesia regarding colonialism's repercussions on Africa's underdevelopment and of an "altruistic" First World's overdevelopment.

Furthermore, *Deliver the Nets* perpetuates an image of Africa that has previously been fixated by photography and mass media but the difference is that now the game *enacts* and emotionally engages us. We actively engage in eliminating malaria rather than "passively" watching a movie and it is particularly important to critically address this relationship between our active engagement and the construction of our emotional attachment to the game. The articulation of simulation and emotions produces such a narrative that we become subjects "to help solve this problem, make poverty and malaria history" but do not question it in historical terms.[3]

In this sense, the esthetic and emotional effect is similar to Brock's (2011) criticism of *Resident Evil 5* of "when keeping it real goes wrong," that is, the representational reservoir called upon works to solidify the essence of Africa "as a geographic and emotional marker for Whiteness" (p. 443). As with Brock's (2011) critique of *Resident Evil 5*, *Deliver the Nets* effaces the political economic and cultural diversity of the African continent in implicit favor for an unambiguously unified image of Africa as a nation (i.e., peoples) in need of being "cleansed and civilized, a role Africa has symbolized to the West for Centuries" (p. 434). The point is not to belabor the burden of representation—as what image could capture the totality of life—but rather to acknowledge the politics embedded within the consistency of this particular representation of Africa. For *Deliver the Nets*, it seems telling that beyond the Avatar—who operates as an extension of the player, much like an indigenous operative does for a humanitarian organization—that the recipients of our philanthropic aid are represented as static images: it suggests that it is our first-world aid that sparks whatever limited agency third-world populations may have to help themselves.

This configuration of first-world mobility and third-world immobility, however, is no accident of game design. It is not that the *Deliver the Nets* developers forgot to mobilize the recipients of philanthropic aid as represented in the game—though clearly this could have been done. Rather, the animation embedded within any representation of third-world subjectivities, by virtue of being nonplayable characters, necessarily requires the scripting of their past, present, and future in advance. For if game design is at once an attempt to emotionally educate and discipline the player according to a particular algorithmic logic (Crogan, 2003), then must it not first make a claim to a particular configuration of the world (Manovich, 2001)? To the extent that game design continues to operate according to first-world logics, in which agency is embedded within the Avatar, then the generic conventions of game design remain encapsulated within what White (1973) identified as a romantic conception of historical processes; that is, romanticism is a historiographical mode of thought that imagines "the

---

[3] This reminds one of the garbage that was left at the end of Live 8 concerts. As people helped make poverty history, they were leaving a huge amount of garbage behind. Furthermore, some of the bands who contributed to the campaign were criticized for flying their private jets to the concerts. On a final note, the symbol of the campaign is ironic in that we see a guitar whose body represents Africa and "8" is symbolized in such a way that it undoubtedly looks like a dollar, standing on the keyboard of the guitar connected right into Africa.

hero's transcendence of the world of experience, his victory over it, and his final liberation from it" (White, 1973, p. 8). Within this mode of historical emplotment, the world is presented as an ahistorical challenge to be overcome—time does not exist until the hero/avatar compels it into existence.

To echo the words of Césaire (1955/2000), some will say but the discourse of the game and website is such that it helps us to solve this problem, and to make poverty and malaria history, but we respond by emphasizing that we are "talking about [...] cultures trampled underfoot, institutions undermined, lands confiscated, [...] extraordinary *possibilities* wiped out" (pp. 42–43). These possibilities will not be realized unless we move beyond a conception of historical time as driven by the will of first-world subjectivities; for as powerful as these subjectivities may be, their mobilization remains contingent upon the immobility of third-world populations, and hence continues the cycle of domination. This is not to suggest that nothing is to be done—for surely something must—but it must be done within a framework of historical time that moves beyond romantic conceptions of historical time as contingent upon the agency of first-world populations. Otherwise empowerment will be limited to the development of powerful agents able to take on the problems of the world, which typically is contingent upon the extraction of resources from third-world populations, thereby continuing the cycle of global inequality. Contemporary gaming practices, whether serious or otherwise, are implicated in this practice and thus remain limited in their ability to pose viable philanthropic alternatives.

CONCLUSION

In the past decade, the serious games movement has garnered enough institutional support to begin demanding for the gamification of a various social problems. Industry conferences, Ted Talks, and research institutes have emerged that are dedicated to the "gamification" of various political, economic, and social problems. As Fishman and Holman (2015, p. 17) argue, focusing on education, "A growing chorus argues that 'gamification' is the answer." This analysis, however, compels us to explore the consequences of this growing call for gamification. For as we have argued, because game design is embedded within the historical totality of its moment of production and ongoing consumption, then game design is necessarily caught up within the neocolonial moment of which it is a part. This effect can be understood as operating on two levels.

First, representations of impoverished populations often reflect the cultural logics of those who create these representations. Hence, it is of little surprise that the continent of Africa is represented as an ahistorical, homogenous whole, as opposed to a collection of disparate countries. The effect is such that *Deliver the Nets* says more about those who created the game—that is Whiteness—then it does about those whom it claims to represent. The broader implications, beyond our specific focus on *Deliver the Nets*, is that game design will be unable to address the problems of our contemporary moment for as long as it remains unable to comprehend the historical totality of our moment; in other words, this is a call for a critical theory of game design.

Second, the substructures of gameplay are contingent upon the very postcolonial conditions of which the serious games movement claims to critique. Or rather, because the serious games movement sees Africa (and other impoverished continents and countries) in terms of a postcolonial framework—as in, after colonialism—the serious games movement is unable to recognize the existence of neocolonial practices that perpetuate the uneven relationship between first- and third-world populations. The effect is that *Deliver the Nets* allows for the player to feel good about their postcolonial contributions while remaining ignorant of their neocolonial practices. More generally, this means that until the serious games movement addresses its neocolonial substructures, in terms of labor, resource extraction, environmental impact, etc. (see Mejia, forthcoming), the movement may well end up perpetuating that which it claims it wishes to resolve.

In conclusion, the serious games movement ought to be understood as an affective rearticulation of what Jodi Dean calls communicative capitalism. These games transform complex social problems into enjoyable experiences that often work to naturalize the primacy of capitalism over alternative political economic systems. Likewise, these games enable us to feel good about our current position in the global economy for these games work to erase our social memory and obscure the neocolonial political economic environment of which we are a part. In sum, the serious games movement requires that we at once appreciate the desire for video games to take up the challenges of our contemporary moment, while simultaneously critiquing the continued reliance upon neocolonial cultural and political economic frameworks.

REFERENCES

Abt, C. (1987). *Serious games*. Lanham, MD: University Press of America.
Arnseth, H. C. (2006). Learning to play or playing to learn – A critical account of the models of communication informing educational research on computer gameplay. *Game Studies*, 6, 1.
Berger, J. (1972). *Ways of seeing*. London: Penguin Books.
Brock, A. (2011). "When keeping it real goes wrong": *Resident Evil 5*, Racial Representation, and Gamers. *Games and Culture*, 6(5), 429–452.

Bulut, E., Mejia, R., & McCarthy, C. (2014). Governance through philitainment: Playing the benevolent subject. *Communication and Critical/Cultural Studies*, 11(4), 342–361. http://dx.doi.org/10.1080/14791420.2014.951948.
Carey, J. W., & Quirk, J. (2009). The mythos of the electronic revolution. In J. W. Carey & G. Stuart Adam (Eds.), *Communication as culture*. New York: Routledge.
Césaire, A. (1955/2000). *Discourse on colonialism*. (J. Pinkham, Trans.). New York: Monthly Review Press.
Crogan, P. (2003). Gametime: History, narrative, and temporality in combat flight simulator 2. In M. J. P. Wolf, B. Perron, M. J. P. Wolf, & B. Perron (Eds.), *The video game theory reader: Vol. 1* (pp. 275–301). New York: Routledge.
Dean, J. (2009). *Democracy and other neoliberal fantasies: Communicative capitalism and left politics*. Durham, NC: Duke University Press.
Dyer-Witheford, N., & de Peuter, G. (2009). *Games of empire*. Minneapolis, MN: University of Minnesota Press.
Eikenberry, A., & Nickel, P. (2006). *Towards a critical social theory of philanthropy in an era of governance*. Blacksburg, VA: Social, Political, Ethical, and Cultural Theory Research E-ditions.
Ewen, S. (1988). *All consuming images: The politics of style in contemporary culture*. New York: Basic Books.
Fisher, E. (2010). *Media and new capitalism in the digital age: The spirit of networks*. New York: Palgrave.
Fishman, B., & Holman, C. (2015). When gamification isn't enough… be gameful. *Spectra*, 51(2), 16–17.
Galloway, A. (2006). *Gaming: Essays on algorithmic culture*. Minneapolis, MN: University of Minnesota Press.
Gee, J. P. (2005). Good video games and good learning. *Phi Kappa Phi Forum* 85, no. 2.
Gee, J. P. (2007). *What video games have to teach us about learning and literacy* (Rev. and updated ed.). New York: Palgrave Macmillan.
Hardt, M., & Weeks, K. (Eds.), (2000). *The Jameson reader*. Oxford: Blackwell.
Kerr, A. (2006). *The business and culture of digital games: Gamework/gameplay*. London: Sage.
Kirriemuir, J., & McFarlane, A. (2004). Literature review in games and learning. *Futurelab series*. Bristol: Graduate School of Education. https://hal.archives-ouvertes.fr/hal-00190453/document.
Manovich, L. (2001). *The language of new media*. Cambridge, MA: MIT Press.
McGonigal, J. (2011). *Reality is broken: Why games make us better and how they can change the world*. New York: Penguin Books.
Mejia, R. (forthcoming). Ecological matters: Rethinking the "magic" of the magic circle. In M. Kapell (Ed.), *Pressing reset: Restarting the conversation about gaming*. Jefferson, NC: McFarland.
Neubauer, R. (2011). Neoliberalism in the information age, or vice versa? Global citizenship, technology, and hegemonic ideology. *Triple C*, 9(2), 195–230.
PBS Frontline. (2010). *Playing America's army*. http://www.pbs.org/wgbh/pages/frontline/digitalnation/waging-war/a-new-generation/playing-americas-army.html.
Rose, G. (2001). *Visual methodologies: An introduction to the interpretation of visual materials*. London: Sage.
Roy, A. (2010). *Poverty capital: Microfinance and the making of development*. New York: Routledge.
Susi, T., Johannesson, M., & Backlund, P. (2007). *Serious games: An overview*. Sweden: University of Skovde.
Tucker, D. (2012). *Gaming our way to a better future*. Washington, DC: Woodrow Wilson Center.
White, H. (1973). *Metahistory: The Historical Imagination in Nineteenth-Century Europe*. Baltimore, MD: John Hopkins University Press.

CHAPTER 11

Police Body Cameras: Emotional Mediation and the Economies of Visuality

Stacy Wood
University of California, Los Angeles, CA, USA

INTRODUCTION

On May 1, 2015 the Justice Department announced a program to fund grants for police departments to purchase body cameras as well as to provide for training and evaluation programs. The twenty-million dollar initiative is part of a larger and longer-term proposal from the Obama administration for seventy-five million dollars of support over the next 3 years. In the wake of the deaths of Michael Brown, Eric Garner, and Freddie Gray and the subsequent movements including "Hands Up, Don't Shoot" and "Black Lives Matter," concerning institutional police violence in communities of color, body-worn cameras have become the palliative of choice from public officials, police departments, and some members of the public. Hillary Clinton has come out as saying that every police department should implement body-worn cameras, the Los Angeles Police Department has begun a pilot program and the Baltimore Police Department has expedited an existent plan for implementation. A statement from Attorney General Loretta Lynch explicitly made body-worn cameras interchangeable with transparency; "Body-worn cameras hold tremendous promise for enhancing transparency, promoting accountability, and advancing public safety for law enforcement officers and the communities they serve" (Berman, 2015). Support for the adoption of body-worn cameras by police officers has certainly created strange bedfellows. Seen as a tool for police accountability by outraged communities and as a preventative measure against unthinkable tragedy by mourning families (Van Wagtendonk, 2014a), they are also seen and have been adopted ad hoc by police officers that consider them a safeguard against false accusation or scrutiny. Body-worn cameras in this sense, then are assumed to exist somehow outside of the exchange or incident itself as a tool for capture, as a neutral, antiemotional, antibiased technological object. The policy rhetoric that frames body-worn cameras not only enters into discourses of surveillance but also into long-standing debates concerning the agential or productive forces of technological objects. This chapter seeks to think through the ways in which the design elements and affordances of body-worn cameras interact with these overlapping discourses and contribute to an understanding of the body-worn camera as a neutral technological mediator of fundamentally emotion-laden human interactions. This understanding, which frames not only the camera itself, but its resulting video evidence as somehow more clear, more easily read, and more neutral, elides the contingency of interactions with the police. State power is intimately tied to visuality both in an esthetic sense and in the ability to see and be seen. The centrality of technologies of visuality to projects of social control foregrounds the importance of control over representational methods. Time and again, citizen documentation and photography become characterized as criminal while the acts they capture do not. This power is bolstered by an investment in the neutrality of technologies of surveillance and is simultaneously reinforced by the depiction of victims of police brutality as emotionally volatile and transgressive. Additionally, any discussion of body-worn cameras must contend with the ways in which different people experience space and surveillance differently, and how they respond emotionally. By introducing the concept of paradoxical space in conjunction with discourses of surveillance and the elements of body camera design, we are able to understand how emotions become a part of an economy of visuality.

POV: BODY-WORN CAMERAS IN THE DISCOURSE(S) OF SURVEILLANCE

Debates in surveillance studies have largely existed between two competing visual metaphors, each reliant on an associated technological apparatus. The panopticon, Jeremy Bentham's design for prisons from 1791 was expanded into a complete and complex representation of power by Michel Foucault (Foucault, 1977) in *Discipline and Punish,* his history of institutional and state power. This visual metaphor maps completely onto the dynamics of power for Foucault, linking the affordances of a central observational tower to the relationship between the seeing and the seen. Part of Foucault's fundamental argument also asks for the internalization of these power dynamics, the production of a citizen who internalizes the concept of total surveillance and acts accordingly. This disciplinary function is aligned with the technology, fundamentally altering not only the relationship between the citizen and the state, but also between the citizen and their own understanding of their subjectivity. The act of surveillance calls attention to itself visually but also abstracts its relationship to human agents, becoming a representation of the state as a political entity. This concept of surveillance is totalizing but unidirectional in this model, and relies on an antagonistic relationship between state and citizen. This model creates a set of emotional relationships through the assumption of the recognizability of surveillance technologies. Symbols then of state power, such as police uniforms or military vehicles, can create a particular emotional state depending on an individual's or a community's history with and expectations of state power. A police vehicle that might be seen and understood as a symbol of safety and reassurance in one neighborhood might be viewed in an entirely different way in another neighborhood that has a legacy of police violence and persistent harassment. The visual regime of panoptic power relies on the visibility both of the surveillance infrastructure and the surveilled themselves, assuming the easy identification of both sides of this dualism. Often understood as an overriding feature of modernity, the panopticon model has been challenged and invoked in various settings, in relation to changing technological and media arrangements. Attempts have been made to extend Foucault's analysis in light of technological change and the proliferation of more intimate and immediate forms of surveillance that come from not only public, but also corporate entities. Mark Poster's notion of the superpanopticon acknowledges the deterritorialization of surveillance technology. While he doesn't necessarily complicate the relationships and dynamics created by these technologies, the superpanopticon recognizes that the surveilled and the technologies of surveillance do not necessarily need to share space, nor do they need the same kind of visual recognizability. His description argues for an even more embedded and ubiquitous form of surveillance that also relies on a different level of complicity. Poster characterizes this form of surveillance as somehow less tied to physical space or order. However, these arrangements are still fundamentally physical and locatable, but resemble the networked architecture that allows their capabilities. Diana Gordon dubbed this phenomenon the electronic panopticon. Both Poster and Gordon insist on maintaining ties to Foucault's analysis of the panopticon because in spite of the new technological capabilities that extend the phenomena of surveillance, they believe that the fundamental dynamics do not shift.

In their 2000 essay "The surveillant assemblage," Kevin D. Haggerty and Richard V. Ericson (Haggerty & Ericson, 2000) base their critique of the panopticon surveillance model in an opposition to foregrounding these changing technological and media arrangements. Steeped in the theoretical and analytical tools developed by Gilles Deleuze and Felix Guattari, their argument focuses on the nature of the assemblage. Assemblages are made up of a multiplicity of objects, subjects, and processes that are characterized collectively only as they work together. Assemblages challenge the assumptions of stability and unity that characterize the panopticon model, stressing that asymmetrical power dynamics are emergent when assemblages themselves become stable or static. By locating the potential for surveillance not in the technologies themselves but in the desire to gather information and the justifications that emerge out of this desire, the assemblage model interrogates the novelty of surveillance capabilities and reminds us of the centrality of in-person observation and interpretation as the mode of bringing together what Roger Clarke referred to as dataveillance (Clarke, 1988). One of the major critiques of the panopticon model of surveillance is its rigidity. Conceptualizing surveillance as contingent and temporary assemblages provides for the possibility of different, radical arrangements. The emergence of counterveillance strategies in both organized and informal settings can be understood as a form of disruption, but they are always disruptions reliant on the logic of surveillance. Enabled by the same networked technologies that precipitate dataveillance, contemporary counterveillance techniques range from encrypted software to using cell phones to film interactions with the police.

When thinking about body-worn cameras as a technology and mode of surveillance, we must contend with not only its rhetorical uses and its design elements or affordances, but also with its status in the discourse of surveillance. How do we characterize the type of surveillance that body-worn cameras enable? They are simultaneously owned and operated by the state and in some vital sense remain unidirectional while they are also expected to be an accountability tool, acknowledging police brutality and misconduct. As such they are also mechanisms of a form of workplace surveillance (Introna, 2000) for police officers themselves. The timing of the call for body-worn cameras also marks their use as a strategy for the maintenance of policing practices and protocols, allowing for an elision of the challenge to propose substantive change to policing or movement to dismantle the criminal justice system. The body-worn camera shifts the onus of transparency and accountability from the police themselves and relocates it in a technological object. In this way, the cameras also operate as tools for controlling the emotional state and outcry of the public. Instead of indictments for individual police officers or strategies for significant structural change, we have a seventy-five million dollar investment in a technological solution that is in and of itself considered enough. It is also necessary to point to a previous technological solution posed during a different historical moment, the proposal of police dash cams arose in a similar political climate in which there was mass public outcry in response to violent interactions between police and communities of color. The dash cam was hailed as a solution to contentious police activities, a new era for transparency and accountability. The difference between a dash cam and the body-worn camera is one of design and implementation but the hopes for their effects are the same.

DESIGN ELEMENTS OF AN ANTIEMOTION

An emotion is a state of mind "deriving from one's circumstances, mood, or relationships with others" (Emotion, 2015). Police often intervene in already emotional circumstances, and alternately can elevate the emotional tenor of a circumstance through their presence. Much work has been done in categorizing and classifying types and states of emotion. Plutchik categorized the basic array of emotions as: acceptance, anger, anticipation, disgust, joy, fear, sadness, and surprise (Plutchk, 1980). As opposed to more persistent affective states such as disposition, mood, or sentiment, emotions can change rapidly depending on circumstance and are recognizable through their physical expression. Emotions not only affect the mind and body at a physical level on behalf of the person having the emotion, but those emotions affect other people through their expression. These expressions enter into a visual economy in which they are more or less easily recognized and read. They can also be exaggerated or downplayed based on particular cultural filters (Ugur, 2013a). These cultural filters point to not only to the existence of shared associations but also marks the root of those shared associations in common identities based on race, class, gender, sexuality, and ability. The testimony of police officers and the coverage by the media exhibit the intimate link between these visual economies of emotional recognition and legal outcomes. Darren Wilson, whose physical descriptions of Michael Brown during his testimony recalled racist language and stereotypes, also zeroed in on his perception of Michael Brown's emotional state. His statements become pattern, Brown had "the most aggressive face," and he "looks like a demon, that's how angry he looked" (Cassidy, 2014). It is often the case in contested incidents of police brutality that the victim's emotional state serves as justification, based on an officer's interpretations and associations.

After Darren Wilson was not indicted for the fatal shooting of Michael Brown, Brown's family released a statement focusing on "fixing the system," and suggesting that the system could be fixed by helping to "ensure that every police officer working the streets in this country wears a body camera" (Van Wagtendonk, 2014b). The rhetorical power of the body camera holds purchase here, simultaneously claiming that it can act to stop tragedy with its mere presence or if tragedy occurs, that it will provide an unbiased record of the incident. The body camera is not an obvious solution, visual materials acting as evidence in the courtroom have been mired in debate and contestation for centuries (Burton et al., 1999; Meskin & Cohen, 2008), it is neither guaranteed nor plausible that the existence of video footage will stand on its own. However, returning to the description of policy discussions that began this paper, it is easy to see why it is a solution that is both emotionally satisfying and tempting. If somehow the interpretation of events was not up to just one police officer, if only we had a video of the incident that led up to the death of Michael Brown, but of course there are numerous cases in which we have had video or we've had partial video. The death of Eric Garner was entirely captured on camera, as were the murders of John Crawford and Oscar Grant (Goodman, 2014). Members of the public shot these videos and the assumption is that the technological capabilities and trustworthiness of those video documents is suspect and open to manipulation. Body-worn cameras are presumed to have features compliant with standards of evidence and chain of custody, policies of accountability that are subject to police practice.

The announcement of President Obama's plan for a seventy-five million dollar investment in police body-worn camera programs seemed an acknowledgement of the problem but was also quickly named as a "windfall" for companies producing the devices (Stone, 2015). These tragic incidents have been a boon for these companies, whose orders and market share have skyrocketed in the past year. These companies have built-in staying power as access to and retention of these videos becomes another product for sale through subscription cloud storage services. In contrast to the complex emotional geographies represented throughout the life cycle of these incidents, body cameras seem simple, direct, unbiased, and unquestionable capitalizing on the expectation that technologies not driven but rather emergent from prioritizing efficiency or improvement. This assumption of unbiased disaffection is manufactured through the design features of body cameras. How are they designed? By whom and for who are they designed? Does this design have an influence on interactions with the police? In the past few years, discussions have emerged concerning the relationship between emotion and wearable technology and of course, the application of this relationship to marketing strategies. Predominantly concerned with the ways wearable technology will capture and respond to emotions of the wearer (Picard, 2010; Ugur, 2013b), these discussions rarely include any commentary or critique of the body-worn camera in the law enforcement context. The design elements and esthetics of body-worn cameras mark them as police equipment rather than consumer product, Google Glass, although technically capable of all necessary functions of the body-worn camera is neither cost-effective nor visually analogous with police equipment. Body-worn cameras are not made equal, and in order to think about the relationship between body-worn cameras and emotion, it is especially important to understand the differences in the elements of design and affordances between the best selling models with law enforcement agencies. Three companies exist at the top of the body-worn camera market, Taser, VIEVU, and Digital Ally.

VIEVU has two models of body-worn camera, one that is marketed as a consumer product. It is "designed for professional consumers who have liability in their lives" (Shu, 2014). It records minimal, lower grade video footage, does not have the capacity for shooting photo, and has capabilities for live streaming. It stores video straight to your iPhone or Android instead of a secure server. The body-worn camera produced by VIEVU specifically for law enforcement agencies is the LE3. VIEVU uses the slogan "designed by cops for cops" (PoliceOne staff, 2012) to emphasize the intimate link between design and use here. The device is an HD camera designed specifically to meet the needs of law enforcement and builds in some measure of evidentiary standards as well. The VIEVU LE3 has a lens cover that functions as an on and off switch for the camera, leaving the discretion of recording up to the individual officer and also has a mute function. It has long battery life and 16GB of nonremovable storage; this model also works in concert with proprietary software, VERIPATROL that allows for the storage of all video evidence to local servers. VIEVU has also developed VIEVU Solution, a hosted cloud-based platform on Microsoft Azure Government. Microsoft Azure Government foregrounds compliance with the standards put forth by the Federal Bureau of Investigation's Criminal Justice Information Services Division. The camera has low light capability, a 68-degree field of view, is lightweight and about the size of a pager. It clips securely to either the chest area or the belt, allowing for seamless integration into the police uniform.

While the slogan "designed by cops for cops" should leave no mystery, the VIEVU website frames its services as on behalf of law enforcement. "The liability present in law enforcement actions creates the need for officers, and administrators, to have protection from any unfounded complaint that may arise. The new technology also provides indisputable evidence for later use in the prosecution of criminal cases" (Vievu, 2015). Neither VIEVU model calls attention to itself, especially in the context of an interaction with the police. Rather than mediating emotion or bolstering a sense of security from the public, the camera is designed specifically to support the officers themselves. VIEVU's previous LE2 model was ranked the highest by a focus group study done by the Department of Homeland Security (U.S. Department of Homeland Security System, 2015). One large study has been done on the effect and use of body-worn cameras inside of law enforcement agencies (Barak et al., 2015) and much of its effect relies on the fact that the cameras worn by police were highly visible. This study contends that both complaints against police and abuse by the police themselves goes down with visible body-worn cameras, suggesting that being watched leads to what they describe as "socially-desirable behavior" or in other words a lack of escalation in interactions with the public.

Taser produces both consumer grade and law enforcement grade products. Their slogan is "Protect Life. Protect Truth," their website (Taser, 2015) declares the slogan over a scrolling banner of photos of police officers carrying children and proudly fastening their Taser AXON body camera to their uniform. Taser makes two body-worn cameras, the AXON Body and the AXON Flex. Aside from the significantly lower cost of the AXON Body, the most distinctive feature of both AXON cameras is the thirty-second buffer feature, which allows for the capture of 30 seconds prior to any recorded incident. The recording is still under the control of the police officer and the buffering is often framed in a way that protects the integrity of the police officer's actions and version of events, "The buffer only saves the video portion and not the audio to protect private conversations prior to the start of the recording. This is vital because when an offender decides to create a situation in which some force option is necessary on the part of the officer, the offender's actions are captured" (Wylie, 2013). The AXON Flex is the only model that places the camera at eye level of the officer. It can be mounted on glasses, a cap or a collar, and is connected to a storage device clipped onto the officer's uniform making it the most explicit and visible of any available model. Taser has also developed its own proprietary software and cloud-based storage service, creatively titled Evidence.com. Plans are subscription based, leaving the fate of these police created public records at the whim of Taser itself.

Digital Ally offers two models of police body cameras. The FirstVu HD comes with a main recorder, body camera clip on module, and a glasses camera. The description on Digital Ally's website stresses that you can "Record video and optional audio from your own point of view wherever you need it, day or night, to protect your job and department, increase the rate and speed of convictions, and more" (Dgitalallyinc, 2015). Here again, the product is explicitly framed as a welcome addition to policing infrastructure rather than fostering the possibility of mediation. The FirstVu HD comes with VuLink, an optional way of setting the camera to automatically record when you turn on your siren or open your door (for example), and works in concert with the Vehicle Video System, a Digital Ally dash cam product. This model also comes with a "Covert Mode" which deactivates the LEDs on the body camera, blending the device in with the police uniform. Additionally their FirstVu Single-Enclosure Body Cam with Monitor comes with similar features in a single unit with no component parts. An added feature is a more robust capture of data to accompany video. This is also the first product description that stresses that "recorded evidence cannot be edited or deleted on the system and is watermarked to prevent tampering" (Digitalallyinc, 2015). The recorded evidence makes its way along a similar chain of custody that other gathered evidence would also follow.

All three manufacturers frame their products as objects that will remove annoyances, complications, and impediments from police work. The esthetics of law enforcement product design and sales is remarkably consistent; the devices all blend into the police uniform so as not to call attention to the device. Designed with the ease of the officer in mind, each device is also imbued with an implicit trust in law enforcement. All devices leave the decision of when to record up to the officer, they leave audio optional so as not to infringe on the privacy of the officer while simultaneously disregarding the privacy or sensitivity of the person interacting with law enforcement. Their criminality is already assumed, and records are then generated, stored, and accessible to law enforcement.

These records depending on their circumstances can be intimate, embarrassing, traumatizing, capturing a vulnerable and exposed moment that now has a visual record. I will narrowly invoke McLuhan here as he conceives of technology as the social, physical, and psychic extensions of the human body (McLuhan, 1964). The body-worn camera does not stand apart as neutral party from law enforcement but is instead an extension of its logics and aims.

PARADOXICAL SPACE AND THE POLITICS OF VISUALITY

In the 1993 book Feminism and Geography, Gillian Rose proposed the term paradoxical space (Rose, 1993) in order to contend with the contingent experience of space. Rose uses the concept in a multiplicity of ways but it is always in service of radical possibilities. Paradoxical space acknowledges the contingency of spatial relationships and experiences, challenging the field of geography to not only think through the ways in which people inhabit spaces differently but to contend with hybrid and intersectional identities. For the study of surveillance, paradoxical space holds significant implications and requires us to think beyond the static dialectic of observer and observed. This concept also requires us to think beyond the persistent discourse of individual privacy. Much of the discourse posing individual privacy against its assumed opposite surveillance stops short of contending with the complex relationships and configurations captured during and precipitated by police interactions. Communities adversely affected by structural racism, institutional violence, and police harassment always already inhabit a paradoxical public space in which not being observed is characterized as a privilege. Communities of color, indigenous people, queer people, and poor people have historically had to negotiate this paradoxical relationship to surveillance and discourses of privacy. Immigrant communities are the first to experience large-scale biometric data collection in the United States, a practice that has recently led to concerns about how data is collected, accessed, stored, and used (Lynch, 2012). Concerns about fraud have let to government data collection policies tied to EBT and other social welfare programs (Robertson, 2000), and of course there are more direct forms of surveillance such as racial profiling and stop-and-frisk policies (Goldstein, 2013). New technologies and forms of surveillance are often implemented in low-income communities, immigrant communities, and communities of color first. Privacy then, is already a privilege, something that you are denied access to if you need any form of government assistance. These video records are another form of this paradoxical space. In the case of Eric Garner, the video was not taken by police, but was filmed by a member of the public. However, it does point to the kinds of negotiation expected from people of color in the face of police brutality. For the promise of safety and possibility of fair treatment, people have to expose traumatic visual materials. Each video begs a series of questions that belie the assumption of clarity that technological solutions tend to carry. Who shot the video? When did it start? Who has access to it? Who gets to speak against or alongside the video? The videos do not show the entire incident either visually or in a temporal sense. These interactions already exist within an economy of visuality and the videos embed these dynamics into an object with evidentiary value. Rather than the word of the police officer, we now have the word of the police officer alongside a video filmed and edited (to relative extent—see product descriptions above) by him that may or may not have audio and that may or may not have started filming at the right moment. I use economy of visuality here instead of visual economy to utilize Hal Foster's definition of visuality as social fact (Foster, 1988). Race, gender, class, ability, and emotion are all marked with visible cues that then enter an economy of visuality that is discursively and historically constituted. Body-worn cameras themselves enter into an economy of visuality that considers technology cold, inert, inhuman, objective, and neutral.

CONCLUSION

The promise of body-worn cameras is a tempting one, a technological solution to a legacy of racialized violence on the part of the police presents itself as simple and uncomplicated, the presence of the camera taking the guess work out interpreting an incident. However, body-worn cameras are manufactured by private corporations and law enforcement is their customer. Although not every company boasts the VIEVU "designed by cops for cops" slogan, each does reveal in their corporate rhetoric that their concerns are for expediting arrests and convictions while protecting individual officers and departments from scrutiny. These cameras hold a unique position in the discourses of surveillance, as reliant on networked technology but also harkening to the disciplinary function of the panopticon model. The emotional registers produced by surveillance are complicated by the asymmetrical level of observation and scrutiny experienced by people of color. Emotions are produced and represented by body-worn camera footage, but their ability to be recognized or understood is not obvious. Race, gender, emotion, and class all exist in overlapping economies of visuality. These economies of visuality are what allow Darren Wilson's description of Michael Brown's demeanor and behavior to retain such purchase in a racist legal system and media landscape. Emotions such as fear and anger become racialized, the videos filmed and edited by the police themselves becoming evidence of their judgement.

REFERENCES

Barak, A., et al. (2015). The effect of police body-worn cameras on use of force and citizens' complaints against the police: A randomized controlled trial. *Journal of Quantitative Criminology, 31*(3), 509–535. Published online November 19.
Berman, M. (2015). *Justice dept. will spend $20 million on police body cameras nationwide*. The Washington Post. Published May 1, 2015, Accessed 12.05.15.
Burton, M., et al. (1999). Face recognition in poor-quality video: Evidence from security surveillance. *Psychological Science, 10*(3), 243–248.
Cassidy, J. (2014). *A closer look at officer Wilson's testimony*. The New Yorker. November 25.
Clarke, R. (1988). Information technology and dataveillance. *Communications of the ACM, 31*(5), 498–512.
https://www.digitalallyinc.com/HD-body-cam.html Accessed 14.05.15.
https://www.digitalallyinc.com/personal-camera.html Accessed 14.05.15.
"emotion, n.". *OED Online*. March 2015. Oxford University Press. http://www.oed.com/view/Entry/61249?rskey=XRXPxz&result=1&isAdvanced=false Accessed 12.05.15.
Foster, H. (1988). Introduction. In H. Foster (Ed.), *Vision and visuality*. New York: The New Press.
Foucault, M. (1997). *Surveiller et punir*, (A. Sheridan, *Discipline and Punish*, Trans.). Harmondsworth, Penguin.
Goldstein, J. (2013). *Judge rejects New York's stop-and-frisk policy*. New York Times. August 12.
Goodman, A. (2014). *No charges in Ohio police killing of John Crawford as Wal-Mart video contradicts 911 caller account*. Democracy Now Transcript. Published September 25, 2014, http://www.democracynow.org/2014/9/25/no_charges_in_ohio_police_killing. Accessed 14.05.15.
Haggerty, K., & Ericson, R. (2000). The surveillant assemblage. *The British journal of sociology, 51*(4), 605–622.
Introna, L. (2000). Workplace surveillance, privacy and distributive justice. *Computers and Society, 30*(4), 33–39.
Lynch, J. (2012). From fingerprints to DNA: Biometric data collection in U.S. immigrant communities and beyond. Report from the Electronic Frontier Foundation and the Immigration Policy Center prepared May 2012.
McLuhan, M. (1964). *Understanding media: The extensions of man*. New York: McGraw-Hill.
Meskin, A., & Cohen, J. (2008). Photographs as evidence. In S. Walden (Ed.), *Photography and philosophy. Essays on the pencil of nature* (pp. 70–90). New York: Wiley-Blackwell.
Picard, R. (2010). Affective computing: From laughter to IEEE. *Affective Computing, IEEE Transactions on, 1*(1), 11–17.
Plutchk, R. (1980). *A general psychoevolutionary theory of emotion*. New York: Academic.
PoliceOne staff. *For cops, by cops: VIEVU body-worn cameras*. Published July 1, 2012, http://www.policeone.com/police-products/for-cops-by-cops/articles/5764127-For-Cops-By-Cops-VIEVU-Body-worn-cameras/. Accessed 20.05.15.
Robertson, R. (2000). Better use of electronic data could result in disqualifying more recipients who traffic benefits. Report from the U.S. Government Accountability Office published March 7, 2000 and publicly released April 6, 2000.

Rose, G. (1993). *Feminism & geography: The limits of geographical knowledge*. Minneapolis: University of Minnesota Press.
Shu, L. *VIEVU2 review: body cams aren't just for cops: The VIEVU2 is a digital witness you can wear*. Published December 5, 2014, http://www.digitaltrends.com/camcorder-reviews/vievu2-review/. Accessed 20.05.15.
Stone, J. (2015). Police body cam manufacturers see windfall in Obama announcement. *International Business Times*. Published December 2, 2014, Accessed 12.05.15, http://www.ibtimes.com/police-body-cam-manufacturers-see-windfall-obama-announcement-1732100.
https://www.taser.com Accessed on 13.05.15.
Ugur, S. (2013a). *Wearing embodied emotions: A practice based design research on wearable technology*. Milan: Springer.
Ugur, S. (2013b). *Wearing embodied emotions: A practice based design research on wearable technology*. Milan: Springer.
U.S. Department of Homeland Security System Assessment and Validation for Emergency Responders Summary on Wearable Camera Systems. https://www.bja.gov/bwc/pdfs/DHS_SAVER_Wearable-Camera-Systems_Comparison.pdf Accessed 16.05.15.
Van Wagtendonk, A. (2014a). *Grand jury won't indict office on Michael Brown death*. PBS NEWSHOUR. Published November 24, 2014, Accessed 14.05.15.
Van Wagtendonk, A. (2014b). *Grand jury won't indict office on Michael Brown death*. PBS NEWSHOUR. Published November 24, 2014, Accessed 14.05.15.
http://www.vievu.com/about-us/ Accessed 16.05.15.
Wylie, D. *New Taser AXON Body on-officer camera hits the streets*. Published August 1, 2013, http://www.policeone.com/police-products/body-cameras/articles/6354361-New-TASER-AXON-Body-on-officer-camera-hits-the-streets/. Accessed 12.05.15.

INDEX

Note: Page numbers followed by *f* indicate figures, *t* indicate tables and *np* indicate footnotes.

A

Affective channel (AC), 132, 134–135f, 165
Analogical reasoning, 103–107, 104f
Anthropomorphism, 32–33, 183–185
Army Battlezone, 215–216
Artificial intelligence, 171–172
  anthropomorphism
    goal of, 183–184
    human's replicability, 184–185
    "program-resistant" functions, 184
  emotions in
    definition of, 173–177
    ELIZA, 180
    higher emotions, 174
    human-computer interaction, 181
    human verbal and nonverbal emotional communication, 180–181
    imitation game, 175–176
    lower emotions, 174
    MIT's Humanoid Robotics Group, 180–182
    MIT's Personal Robots Group, 182
    multitask system, 182
    neurology and cognitive science, 173–174
    nonhuman processes, 182–183
    statistical *vs.* simulation mode, strong *vs.* weak artificial intelligence, 177–179
    techno-handshake, 180
Auditory display
  Apple iPhones and smartphones, 29–30
  auditory feedback, 29–30
  definition, 29
  digital security systems, 29–30
  earcons, 30
  visual icons, 29–30
AXON cameras, 234–235

B

Biodata, 206–207
Black-and-white, 13

Body-worn cameras
  agential/productive forces, 227–228
  emotion, design elements
    AXON cameras, 234–235
    FirstVu HD, 235
    Google Glass, 232–233
    HD camera, 233–234
    law enforcement context, 232–233
    marketing strategies, 232–233
    VERIPATROL, 233–234
    VIEVU, 233–234
    VIEVU LE3, 233–234
    visual economy, 231–232
  paradoxical space and visuality, 236–237
  surveillance studies
    dash cam, 230–231
    dataveillance, 230
    Foucault's analysis, 228–229
    Foucault's fundamental argument, 228–229
    panopticon model, 228–230
    state and citizen, 228–229
    training and evaluation programs, 227–228

C

Cheerful strategy, 137
Cinematic emotion
  artificial and self-referential color patterns, 16
  audio-visual products, 6
  brain stimulation, 7
  color, effect of, 4–5
  color factor, 3
  color memory, 7
  color perception, 7
  color selection, 8
  color stimuli, 8
  contemporary cinema, 3–4
  cultural structure, 16
  digital cinema and video, 4
  distracted perception, 8

Cinematic emotion *(Continued)*
   emotional pattern
      black-and-white, 13
      chromolithography, 10
      color design, 9–10
      color's sensory impact, 9–10
      cultural stereotypes, 14–15
      digitalization of cinema, 9–10
      epidermal stimulation, 14–15
      female models, 11–12
      Gucci Bright Diamante, 11
      media products, 9
      Renault Clio Costume National, 12
      saturated and spot colors, 11–12
      yellow complexion, 14–15
   Hollywood classical cinema, 5
   marketing and advertising, 8
   monodimensional perceptual experiences, 6–7
   neuroscience, 3
   production and consumption, 5–6
   psychological and behavioral significance, 5
   reproductions of paintings/film, 6–7
   Technicolor, 5
   viewer, cognitive activity of, 15
   visual culture, 8–9
Cognitive load theory, 109–110, 116
   extraneous cognitive load, 110
   germane cognitive load, 110
   intrinsic cognitive load, 110
Companions
   affective channel, 132, 134–135f
   children and companionship, 130
   children's perception, 142
      focus group activities, 151–154, 155t
      interview sessions, 154–155, 156t
   CIS, 139t
      confused strategy, 138
      idle/stand-by strategy, 138
      interrupted strategy, 138
      listening strategy, 138
      surprised strategy, 137
   collaborative co-located learning game, 150, 152f
   collaborative learning interaction strategies, 138, 139t
   control room and experiment room, 150, 151f
   definition, 129–130
   designing interaction strategies for backchannels, 134–135
   face-to-face oral communication, 134
   facial expressions, 132–134
   HWYD companion, 135
   NLU model, 135
   dialogue scripts and sentence patterns, 149t
   EMOTE Project, 131
   emotional interaction strategies, 139t
      cheerful strategy, 137
      inquisitive strategy, 137
      sympathetic strategy, 137
   focus groups and personal interviews, 149, 150t
   Huggable, robotic research, 131
   iCat robot, 131
   interaction with, 159
   introductory dialogue scenario, 153t
   LIREC Project, 131
   mathematics video game, 145f
   Mixtec codices, 144f
   personality descriptors, 156–157, 157f
   physical descriptors, 157–158, 158f
   pilot session, 148–149
   PlayPhysics, 143–144
   positive emotions, 163–164
   preferred activities, 160
   prEmo icon, 162f
   Quest Atlantis Project, 143
   screen-based companion, 164
   serious game, 143
   significance of, 161, 161f
   trust in, 161
   utility/functionality descriptors, 158–159
   "within-subjects" experimental design, 142–143
   WoZ system (*see* Wizard of Oz (WoZ))
Confused strategy, 138
Control board (CB), 147–148, 147f
Conversational interaction strategies (CIS), 139t
   confused strategy, 138
   idle/stand-by strategy, 138
   interrupted strategy, 138

listening strategy, 138
surprised strategy, 137
CyGaMEs
    chunks information, knowledge schemata, 111–112
    Flowometer Report, 119, 120f, 121
    instructional digital game systems, 106f
    instructional science game, 121
    knowledge-learning-integration framework, 108–109, 108f
    knowledge structures, 111–112
    nine-channel model, 119, 120f
    Timed Report, 118–119

D

Deliver the Net, 214–215
    Berger's analysis, 220
    campaign, impact of, 218
    communicative capitalism, 224
    emotional cultivation, 219
    game design, 219, 222–223
    games movement, 224
    historical moment, 220
    philanthropic aid, 222–223
    photography and mass media, 221–222
    player role, 218–219
    political economic and cultural diversity, 222
    postcolonial conditions, 224
    romantic conception, of historical processes, 222–223
Detection and Analysis of Psychological Signals (DCAPS), 83–84

E

Edutainment, 143
Embodied conversational agents (ECAs), 129, 133f
EMOTE Project, 131
Emoticons
    communications, 38
    computer-mediated-communication, 37–38
    Eastern emoticons, 43t
    emoticons resemble facial nonverbal behavior, 44
    face-to-face (FTF) communication, 37–38
    facial *kaomoji*, 43–44
    and gender, 45
    and icons, 44
    illocutionary force, 44
    LMX (*see* Leader-member exchange (LMX))
    messages, 44–45
    nonverbal cues, 38
    online multiuser dungeon (MUD) games, 45
    professionalism, 38
    sideways emoticons, 43–44
    and smilies, 42–43
    types, 42–43
    Western emoticons, 43t
Emotional interaction strategies, 139t
    cheerful strategy, 137
    inquisitive strategy, 137
    sympathetic strategy, 137
Emotion interface process (EIP), 132
Emotions
    cinematic emotion (*see* Cinematic emotion)
    communicative capitalism, 213–214
    serious games
        Army Battlezone, 215–216
        Deliver the Net (*see Deliver the Net*)
        education, 215–216
Empathetic technology
    altruistic and unselfish behavior, 57
    analytical and empirical methods, 59
    analytic approach, 59
    artificial intelligence, 59
    cohesion and social structures, 57
    definition, 55–56
    empathetic virtual companions
        AIBO, 74
        Aibo, 72
        Clippy, Microsoft Office virtual assistant, 65, 65f
        denotative and connotative meaning, 69
        digital experience, 75
        Enrica, 73
        feature, 63–64
        feeling of security and well-being, 63
        GPS systems, 67
        Hal, 73

Empathetic technology *(Continued)*
  human strategies, 75
  IBO, robot dog, by Sony, 66, 66*f*
  Ifboot, 72
  interaction strategy, 76–77
  Maslow's hierarchy of social needs, 64*f*
  materiality and affect, 71, 71*f*
  Nabaztag, 72
  Nikes with iPod, 73
  Patachon, 73
  pattern grid for experiment, 72*f*
  personal construct theory (PCT), 69
  Pivo, 73
  proactive involvement, 75
  repertory grid, 69
  Samuela, 67*f*, 72
  semantic differential, 69
  semiotic square, 70–71, 70*f*
  sense-making, 69
  social relationships, 63
  task-oriented agents, 63
empirical approaches, 60
experiences and feelings, 55–56
human attributes, 57
human computer interaction
  community, 58
human manifestations, 58
mixed emotions, 62*f*
modern cinema, 57
moral reasoning and pro-social
  behavior, 56
philosophical and religious
  underpinnings, 56
Plutchik's wheel for modeling emotions,
  61*f*
psychological construct, 55–56
Russell dimensional model of
  emotions, 61*f*
social and emotional situations, 55–56
social situation, 58–59

F
Face-to-face (FTF) communication, 37–38
Facial expressions
  designing interaction strategies, for
    Companions, 132–134
  nonverbal communication, 39

  and positive psychology, 96–100
  Wizard of Oz (WoZ)
    verbal statements, 140, 142*t*
    virtual companion, 140, 141*f*
FirstVu HD, 235
Flowometer Report, 119, 120*f*, 121

G
Gestalt principle, 107
Global positioning systems (GPS) navigation
  systems, 32–33, 67
Google Glass
  analysis, 203–205
  biodata, 206–207
  controversial aspect, 188–190
  emotion and resistance
    Android platform, 202
    "Don't be a Glasshole" project, 197*f*
    Glass application, 197
    Information Inequalities, 201
    Kodak's instant photography,
      198–199
    Mann's digital eyewear and
      device, 198
    violence, 198
  emotion, design elements, 232–233
  emotion-recognition software,
    205–206
  facial recognition and surveillance,
    205–206
  fundamental design flaw, 187–188
  future aspects, 208–210
  hands-free, 188
  illegal infringements, 188–190
  promotional video, 189*f*
  San Francisco and glass and class rage,
    190–193
  screenshot of frame, 189*f*
  space and spatiality of life, 207
  surveillance and emotion
    and facial recognition software, 196
    Internet-based reproduction and
      dissemination, 193–194
    panoptic power, 195
    paramount concern, 196
    and physical health, 193–194
    self-surveillance, 193–194

technological intervention, 194–195
US-VISIT program, 196
technological projects, 188
wearable control, 202–203

H

HD camera, 233–234
How Was Your Day (HWYD) companion, 135
Human Brain Project, 171–172
Human computer interaction (HCI), 113–116
Human-machine interaction, 181

I

Inquisitive strategy, 137
Interrupted strategy, 138

L

Leader-member exchange (LMX), 38
  CMC, 49
  communication, 46–48
  correlational studies, 49
  definition, 46
  dimension, 47t
  Email and instant messaging communication, 49
  interactional variables and contextual variables, 46
  leader characteristics, 46
  meta-analysis, 48–49
  nonverbal behaviors, 46–48
  subordinate characteristics, 46
LIREC Project, 131
Listening strategy, 138

M

Multisensory effects
  analogical reasoning
    CyGaMEs approach (see CyGaMEs)
    Gestalt principle, 107
    impact of, 103–104, 104f
    metaphor-enhanced design approaches, 105–107
  cognitive load considerations
    chunking and schemata, learning, 110–113
    extraneous cognitive load, 110
    germane cognitive load, 110
    intrinsic cognitive load, 110
    noise, attention and prior knowledge, 109
  cross-modality interaction, 113–114
  design and assessment, 108–109
  innate perceptual mechanisms
    audition and tactile dimensions, 114–115, 115t
    preattentive vision and haptics dimensions, 114–115, 115t
  interest and engagement, 121–122
  rational design guidelines, 116–117
    learning and embedded feedback/assessment, 118–119
    visual, auditory and haptic/tactile multisensory stimuli, 117–118
  research and development, 122–123

N

Natural language user (NLU) model, 135
Nonverbal communication
  CMC, 39–41
  emotions and attitudes, 39
  facial expressions, 39
  F2F situation, 40–41
  leadership impact, 41–42
  social information processing theory, 39–40
  transformational leadership, 41–42
  workplace and leadership, 40
Novint® Falcon, 117–118

P

Perceptual organization, 114
PlayPhysics, 143–144
Police body cameras. *See* Body-worn cameras
Professionalism, 38

Q

Quest Atlantis Project, 143

R

Resident Evil 5, 222

S

Sound
  ambient sound, 31–32
  ambient sound and environmental noise, 31
  American Speech-Language-Hearing Association, 21
  auditory discrimination, 22
  auditory display, 29
  auditory icons, 29–30
  auditory pattern recognition, 22
  auditory performance, 22
  aural imagination, 19–20
  BMWs, 19–20
  causal listening, 25
  concrete music, 24np
  culture and values, 24
  design, in film, 31–32
  emotional impact, 23
  emotional responses, 19
  emotional robot-human interactions, 32–33
  "extra" sound, 32–33
  filmmakers add extra sound effects, 31–32
  Global positioning systems (GPS) navigation systems, 32–33
  HCI and digital media, 28
  human hearing, 21
  listener's ability, 21
  listening, 23–24
  music psychology, 24
  nostalgic feelings and tastes, 33–34
  psychological and philosophical listening mode, 25–26
  reduced listening, 25
  semantic listening, 25
  skeuomorph, 29–30, 29np
  Skype calls, 32
  sonification, 29
  sound localization and lateralization, 22
  speech, 31
  temporal aspects, of audition, 22
  UX design, 20–21
  visual icons, 29–30
  visual imagination, 19–20
  Windows, "tada" fanfare, 30–31
Spoken dialogue agent system
  brain-wave measuring equipment, 100–101
  DARPA, 83–84
  DCAPS, 83–84
  ELIZA, 84
  expressive facial expressions and positive psychology, 96–100
  facial and word expression, agent's replies
    Bot3D Engine, 86
    content presentation, 87
    emotion-arousing scenarios, 86
    impression evaluation, 87
    impression evaluation task, 86
    user's attributes, 87
    user's "fright" scenario, 88
  hikikomori/social withdrawal, 83
  information and communication technologies (ICTs), 83
  learning customer services, 93–96
  Media Equation, 84
  speech recognition and synthesis system, 100–101
  virtual agent and robot
    animation contents, 89
    content presentation, 90
    debriefing, 91
    emotional, 92
    emotion arousal and controlling emotional valence, 90
    emotionless, 92
    joyful and emotionless faces, 89
    pictorial figures, 90
    user's attributes, 89
Sympathetic strategy, 137

T

Timed Report, 118–119, 121

U

United States Visitor and Immigrant Status Indicator Technology (US-VISIT) program, 196
User emotional state (UES), 132

V
VERIPATROL, 233–234
VIEVU LE3, 233–234

W
Wizard of Oz (WoZ), 165
  control board, 147–148, 147*f*
  evaluation session experiment set up, 145, 146*f*
  facial expressions
    verbal statements, 140, 142*t*
    virtual companion, 140, 141*f*

Printed in the United States
By Bookmasters